THE GREAT HOUSING EXPERIMENT

Volume 24, URBAN AFFAIRS ANNUAL REVIEWS

THE GREAT HOUSING EXPERIMENT

Edited by

JOSEPH FRIEDMAN

and

DANIEL H. WEINBERG

Volume 24, URBAN AFFAIRS ANNUAL REVIEWS

SAGE PUBLICATIONS
Beverly Hills / London / New Delhi

For information address:

SAGE Publications, Inc.
275 South Beverly Drive
Beverly Hills, California 90212

SAGE Publications India Pvt. Ltd.
C-236 Defence Colony
New Delhi 110 024, India

SAGE Publications Ltd
28 Banner Street
London EC1Y 8QE, England

Printed in the United States of America

Library of Congress Cataloging in Publication Data

Main entry under title:

The great housing experiment.

Bibliography: p.
1. Housing policy—United States—Addresses, essays lectures. 2. Housing subsidies—United States—Addresses, essays, lectures. I. Friedman, Joseph, 1944-
II. Weinberg, Daniel H. III. Series
HT108.U7 vol. 24 [HD7293] 307.7'6s [363.5'8] 83-3062
ISBN 0-8039-1991-3

FIRST PRINTING

Contents

Preface

☐ THIS BOOK presents evidence on one of the most important government policy issues of the 1980s—how the government should provide housing assistance to needy families. Both President Reagan's Commission on Housing and the U.S. Department of Housing and Urban Development (HUD) have endorsed the concept of housing allowances (sometimes called housing vouchers) as a major element of U.S. housing policy. The research carried out as part of the Experimental Housing Allowance Program (EHAP) by the authors of this volume provides the best direct evidence on the likely outcomes of such a policy.

EHAP was the result of a merger of two novel elements—housing allowances and social experimentation. The country's experience with conventional housing programs since the 1930s had convinced a growing number of politicians and housing analysts that low-income families in need of decent housing might be better served if they received housing allowances—direct cash payments for housing—than if they were offered subsidized housing built by or for the government. This demand-side, consumer-oriented approach is in sharp contrast to the previously pursued strategy of new construction, carried out in such programs as the Public Housing program, dating from the Housing Act of 1937. A properly designed social experiment, following to the best extent practicable a standard experimental design with "treatment" and "control" groups and careful monitoring of results, was considered an efficient way to test such a new social program without the cost and disruption involved in full-scale implementation of an untested housing strategy.

EHAP was authorized in the early 1970s to allow HUD to evaluate the likely results of a housing allowance approach to housing assistance. EHAP consisted of three major experiments:

- the *Demand Experiment*, whose goal was to examine individual household responses to housing allowances;
- the *Supply Experiment*, intended to determine marketwide effects of housing allowances on housing prices and supply; and
- the *Administrative Agency Experiment*, focused on the behavior of local administrative agencies in running an allowance program.

The experiments began in 1973, and final evaluation was completed in 1982. EHAP involved more than 25,000 households in 12 metropolitan areas, at a cost of almost $200 million.

This volume includes research carried out by researchers at Abt Associates Inc. (contractor for the Demand and Administrative Agency Experiments) and at the Rand Corporation (contractor for the Supply Experiment), as well as commentary by researchers at the Urban Institute (contractor for the "integrated analysis" of the EHAP findings) and a former visiting scholar at HUD charged with review of some of the findings.

The book is organized in five parts. Part I presents a history and overview of EHAP and summarizes the findings of each of the three experiments. Part II discusses household eligibility and participation from three perspectives—random selection, open enrollment, and the role of agency services. Part III examines housing consumption outcomes, including price and supply effects of a marketwide program. Part IV examines a wide variety of other issues—housing search and residential mobility, economic and racial/ethnic concentration, and neighborhood change. Finally, Part V presents three papers that discuss some of the policy implications of EHAP.

We believe the book to be timely and of importance not only to housing policymakers at the national level, but also to local housing officials, housing analysts, urban planners, regional scientists, and housing economists. The book should prove of interest as well to those interested in the results of one of the largest social experiments ever conducted.

This book is the result of over a decade of work by well over one hundred researchers, so it is impossible to identify all the individuals who contributed. Without doubt, though, a major portion of the credit belongs to the Office of Policy Development and Research at the U.S. Department of Housing and Urban Development, and more specifically to Jerry Fitts and Terry Connell of that office.

The success of the research effort owes a good deal as well to the project directors of the three experiments (who are also authors in this book)—William L. Hamilton and Stephen D. Kennedy of Abt Associates Inc. and Ira S. Lowry of the Rand Corporation. Without their leadership, none of these papers would have been written. Credit must go as well to a large number of researchers, programmers, questionnaire designers, field operatives, coders, and secretaries who remain the "unknown soldiers" of the huge effort that became known as EHAP—the Experimental Housing Allowance Program.

Both editors planned and collaboratively reviewed the submissions. Weinberg had the burden of and should take the credit for coordinating the effort, handling the administrative duties, brow-beating the authors, and reviewing the galleys.

Even though the analyses in this book were supported by HUD, and one of the editors is now at the U.S. Department of Health and Human Services (HHS), this book was written by the authors in their private capacities. No official support or endorsement by HUD, HHS, or any of the contractors involved in carrying out the research is intended or should be inferred.

Part I

Introduction

☐ CHAPTER 1, by Joseph Friedman and Daniel Weinberg, presents a history and overview of EHAP. The idea of helping poor families in need of housing assistance by providing them with housing vouchers is traced from its introduction (but rejection) during the debates that led to the U.S. Housing Act of 1937 until the early 1970s. At that time, the disadvantages of conventional housing programs became increasingly apparent, and housing allowances began to be considered a viable policy alternative. (It is worth noting that the current administration has proposed to Congress that housing allowances become the major U.S. housing assistance strategy.) Even in the early 1970s, considerable doubt about a housing allowance-based housing program's effect remained, and it was decided to conduct a series of experiments. The chapter summarizes the basic questions to be addressed by each of the three experiments that made up EHAP.

Chapter 2, by Ira Lowry, principal investigator of HASE, summarizes the findings of the supply experiment. That experiment was designed to address the issue of the magnitude of the housing market response to a large-scale housing allowance program. Three issues were of particular concern to policymakers: (1) Would a housing allowance program cause significant housing price inflation? (2) Would a housing allowance program provide sufficient stimulation to cause an increased supply of adequate housing (through repairs, rehabilitation, and new construction)? (3) Would recipients change neighborhoods? Later, additional questions about recipient response were posed.

Lowry presents an extensive set of findings. Briefly, the findings relating to the initial questions were that (1) a full-scale open enrollment allowance program had no effect on rents or property values, (2) the experimental allowance program did not measurably disturb the housing markets of the experimental sites, and (3) the program had little effect on the physical appearance or social composition of residential neighborhoods. Other important findings are summarized in the chapter.

Chapter 3, by Stephen Kennedy, principal investigator of HADE, summarizes the findings of the demand experiment. Criticism of the housing allowance idea often focused on the presumed inability of low-income households to use their allowances effectively in competition for adequate housing units in the marketplace. In particular, doubts had been raised about the ability of poor households to search for adequate housing and about their ability to negotiate successfully with landlords. The demand experiment was designed to test, using randomly assigned control and experimental households, the ways such households use their allowances.

HADE researchers found that many low-income households were effective in finding adequate housing, though the successful participants were often using the allowance to reduce the out-of-pocket cost of the unit they were occupying initially. However, a number of families offered allowances did not participate (find adequate housing). Housing allowances are an efficient way of providing housing subsidies (more so than most, if not all, supply-oriented programs), but they apparently do not serve the entire population at risk because the housing requirements effectively exclude many of the households initially in substandard housing.

Chapter 4, by William Hamilton, principal investigator of the AAE, summarizes the findings of the Administrative Agency Experiment. A housing allowance assistance strategy represented a considerable change from the conventional housing programs administered in 1970 by HUD. Given the local housing agencies' difficulties with administering its traditional programs, the department believed it needed to demonstrate that existing public agencies could adequately administer a program of housing allowances.

Hamilton emphasizes the "naturalistic design" of the AAE—agencies chose and designed their own procedures rather than implementing a uniform national design. All eight public agencies were successful in designing and operating the program, in identifying and enrolling beneficiaries, and in paying housing allowances. The central theme of the research findings is that alternate methods exist for carrying out administrative functions and that any choice among the alternatives makes a difference for recipient outcomes.

1

History and Overview

JOSEPH FRIEDMAN and DANIEL H. WEINBERG

THE EVOLUTION OF HOUSING PROGRAMS: FROM SUPPLY TO DEMAND SUBSIDIES

In the nineteenth century, public concern was concentrated on sanitation rather than directly on housing, and public intervention in the housing sector was limited to regulatory measures such as building and occupancy codes to assure minimum standards of safety and health. As cities became increasingly populated by poor immigrants, public officials became aware of the severe health problems in densely settled neighborhoods. They proceeded to install or improve water supply and sewage systems. This trend was reinforced in the late nineteenth century as the germ theory of disease gained prominence (Burns and Grebler, 1977: 75). Government requirements were extended to include sanitary and running water systems and improved ventilation in newly constructed buildings.

Housing conditions that may facilitate the spread of disease have been eliminated in the United States. Nevertheless, continuing large-scale public intervention in the housing sector demonstrates that there are additional reasons for public involvement. The two major contemporary rationales for government intervention in the housing sector are "merit goods" and housing market imperfections.

The basic argument for providing cash assistance to households to be used specifically for housing instead of general cash transfers is based on the belief that housing is a "merit good—a form of consumption which society views more important than allowed for by individual choices" (Musgrave, 1976: 215), coupled with the feeling that "the poorest people among us

---

AUTHORS' NOTE: *This chapter is based in part on descriptions of the Experimental Housing Allowance Program by Allen, Fitts, and Glatt (1981); Frieden (1980), Hamilton (1979), Struyk and Bendick (1980), and U.S. Department of Housing and Urban Development (1980).*

should live in better housing than they are able to afford, and that they should be assisted to do so" (Weicher, 1976: 182). In particular, the imposition of building codes to improve the quality of housing increases its cost and places a floor under the rents of such housing. This rent is typically higher than the share of income families at the lower end of the income scale would freely choose to allocate for housing. Housing subsidies could then be viewed as a way to compensate poor families for the government's actions in limiting their housing choices.

Imperfections in the housing market arise from "such factors as the fixed supply of land in urban settings, zoning and discrimination which reduce tenant mobility, linkages between location and job availability, lumpiness of housing outlays, credit risks, and so forth" (Musgrave, 1976: 215). Further, the existence of externalities in neighborhoods whereby the attributes and living conditions of one dwelling unit can affect the value and living conditions of another also argues for the government to attempt structural remedies for these imperfections.

The Housing Act of 1937 proclaimed the goal of decent, safe, and sanitary dwellings for low-income families. During the four decades that followed, various program approaches for meeting that policy goal have been either tried or debated. Beginning with the Public Housing program, which was established by the 1937 Act, these programs were usually directed at the production or supply of housing, with the subsidy permanently linked to housing units, not to families. As long as families remained eligible and lived in these particular units, subsidy benefits were passed on in the form of below-market rents.

In contrast, housing allowances (also called housing vouchers or rent certificates) link the subsidies to families and are directed toward stimulating demand for decent housing. Housing allowances are payments made directly to eligible households to help them pay the costs of living in housing of their choice. Such payments can vary, depending on what income limits or other household characteristics are used to establish eligibility; whether or not households receiving payments are required to live in housing that meets specific quality standards and, if so, the nature of those standards; and the method used to compute the size of the payment received by each household.

Housing allowances are not a new idea. The topic has been debated in Congress in some way for more than 40 years. Housing vouchers were considered but rejected in formulating the Housing Act of 1937. The favored approach was the public housing program that attempted to satisfy several objectives. These objectives included the housing-oriented goals of slum clearance and improved living conditions of low-income people, as well as anti-Depression goals of job creation and income generation through public spending.

The alternative of housing allowances or rent certificates was proposed by organizations like the U.S. Chamber of Commerce, the National Association of Real Estate Boards, and conservatives in both the House and Senate. Proponents of rent certificates argued that the program would operate through the private market and keep government out of the housing business, that it would be more manageable and less costly than building new housing, and that it would grant low-income households more choice.

On the other side of the debate, two of the nation's leading housing experts of that time—Catherine Bauer and Edith Elmer Wood—opposed rent certificates. Bauer argued that they would be administratively unworkable, while Wood held that such a program would not result in any additional housing being built for low-income families. The Senate Committee on Education and Labor finally concluded:

> In dealing with the housing of families of low income, systematic [construction of] low-rent housing should be substituted for [financial] relief [including rent certificates]. This procedure will be cheaper for the government, more beneficial to business, and infinitely more desirable to those of our citizens who are now living in slums and blighted areas, both in urban and rural parts of the country [Semer et al., 1976: 95].

Thus, Congress opted for a housing production approach, and the rent certificate approach was shelved.

Housing allowances were reconsidered in the Taft Subcommittee hearings on postwar housing policy in 1944. However, the final report of the subcommittee stated:

> It has been argued that families should be assisted by rent certificates just as grocery stamps have been furnished to needy families. The number of families entitled to rent certificates upon any such basis would be infinitely larger than those requiring other relief. It is not at all certain that such a plan would bring about improvements in the bad housing accommodations that now exist. In fact, the scheme might work to maintain the profitability of slum areas and, consequently, to retard their elimination. It would certainly require a detailed regulation of private rental quarters both as to condition and rent [Semer et al., 1976: 40].

The rent certificates idea was considered and rejected again in designing the Housing Act of 1949 and in the 1953 report by the President's Advisory Committee on Government Housing Policies and Programs.

A step toward a program of housing allowances was made in the Housing and Urban Development Act of 1965, which added two programs—the rent supplement program and the Section 23 leased housing program. Both moved in the direction of the housing allowance by giving participants more flexibility in choosing the places they could live and by making the value of the subsidy depend on a family's income. The Section 23 housing program allowed public housing authorities (PHAs) to lease existing private dwelling units and subsidize low-income households to live in them. One potential expansion of this program was to provide housing allowances—a direct payment to enable eligible households to purchase adequate housing in the private market. The households themselves would then be responsible for finding apartments and negotiating with landlords. One further step toward such an unrestricted system of housing subsidies was the 1974 revision of the Section 23 program, retitled Section 8. The revision took place after the initiation of the Experimental Housing Allowance Program and was obviously influenced by many features of the experiment's design. The Existing Housing portion of the Section 8 program focused on households as objects of the subsidy and permitted them to locate suitable units by themselves, but the government retained some control over location and assisted in lease negotiation, with payment going directly to the landlords.

During the 1960s, economists, housing specialists, and the public at large criticized most housing programs as inequitable and wasteful—inequitable because they served only a small fraction of the population eligible for housing assistance, and wasteful because the subsidized units were more expensive to build and maintain than similar units in the private housing market. Economists added an efficiency argument: that the housing subsidy received was as an in-kind benefit and was therefore valued by the recipients as worth less than its cost to the government.

The shift to demand-oriented programs was a response to widespread criticism of the production-oriented programs as either too costly, not working as desired, or not serving enough eligible families. These traditional programs had only a limited impact because they required deep subsidies. The Congressional Budget Office (1979) estimated that the cost of supply-oriented programs in 1980 dollars ranged from $2200 to 2530 per dwelling unit per year.

Because the cost per unit was so high, only a limited number of housing units could be built with the available housing funds each year. Consequently, only a small number of income-eligible households could be helped each year. Indeed, in recent years, less than 10% of income-eligible households were receiving housing assistance by occupying subsidized housing. The remaining often equally deserving 90% received nothing. National housing policy had thus created inequities within the low-income population by providing large and costly subsidies to relatively few and nothing to the rest.

A related problem is the high cost of construction programs relative to programs utilizing the existing housing stock. Mayo et al. (1980a, 1980b) estimated that the total annual program cost, including administration costs, required to provide a minimum standard unit by a new construction program was from 35% to 91% higher than the cost required to rent such a unit in the housing market. This much higher cost associated with new construction programs reflects construction, operation, and implementation inefficiencies.

Another problem with construction programs is that they restrict the locational choices of participants and tend to create high concentrations of the poor in racially segregated neighborhoods. The resulting problems were graphically illustrated by the colossal failure of the Pruitt-Igoe Public Housing Project in St. Louis.[1] Although the project should be considered an atypical extreme, its demolition in 1972 convinced many housing experts that a new approach to solving the housing problem was needed.

The research at the New York City Rand Institute lent some additional support to the housing allowance idea. Ira S. Lowry and his associates (Lowry, 1971) analyzed the New York City housing market during the 1960s. They discovered that from year to year, a large volume of sound housing was deteriorating in quality, and more than 30,000 units each year were being taken off the market through demolition, conversion to nonresidential use, or outright abandonment. Between 1965 and 1968, housing losses were greater than new construction by a substantial margin. The main reason for this rising volume of deterioration and abandonment was that a large number of the city's low-income population were unable to pay enough rent to cover the rising costs of operating and maintaining rental property. Landlords who were unable to earn a competitive return were cutting back on maintenance, and, in time, walking away from their buildings.

Lowry concluded from the Rand studies that the most effective way to meet the housing needs of low-income families in New York City was to raise the level of maintenance in existing buildings while they were still in good condition. He estimated that an annual rent increase of $400-$700 (in 1967 dollars) would be required to support moderate renovation and good maintenance in typical older apartments. Since low-income families cannot afford even these relatively small increases, Lowry proposed a housing allowance plan that would augment family income at an average cost of little more than $600 per family.

In 1968 the President's Committee on Urban Housing (the Kaiser Committee) had a mandate to analyze existing housing programs and their impacts on households and markets, focusing on low-income households. The committee's 1968 report argued in favor of a housing allowance and recommended that the government undertake an experiment to determine whether a housing allowance program would be feasible and worthwhile.

The Kaiser Committee's support of housing allowances stemmed largely from the increasing difficulty of the public housing program in finding decent sites for large housing projects and a concern for the consequence of segregating the poor and minorities in limited areas of cities. The committee also stressed that an

> allowance system offers the opportunity for the free market to operate in its traditional fashion [and that] widespread distribution of housing allowances to

> poor families should reduce the economic dependence on slum housing and shift the demand upward for standard units. In response to this shift in demand, suppliers of housing would be induced to produce more standard housing, either by upgrading slum properties or through new construction [p. 11].

The committee recommended not a full-scale national program but rather an experimental one because of several concerns. First, the committee perceived a need in the short term to stimulate new construction and felt that the conventional project subsidy approach would best accomplish that result. Second, the committee was concerned that a massive allowance program "would be likely to inflate the costs of existing housing considerably, at least in the short run. . . . Consequently, any large-scale housing allowance system would have to be introduced gradually." Finally, the committee feared that without "strong programs of consumer education and vigorous attacks on racial discrimination" an allowance system could have adverse results (p. 71).

The publication of the Kaiser Committee's recommendations coincided with the change from the Democratic administration of Lyndon Johnson to the Republican administration of Richard Nixon. Incoming officials in the Office of Management and Budget and the Department of Housing and Urban Development (HUD) were interested in the idea of housing allowance as consistent with Republican goals of increased utilization of the private market in public programs (Struyk and Bendick, 1981: 29). In 1970 the first housing allowance demonstration program (under the auspices of the Model Cities program) was launched in Kansas City, Missouri and in Wilmington, Delaware. HUD began preliminary studies and designs for a systematic national experiment in 1970 and 1971, and then organized its Experimental Housing Allowance Program (EHAP).

In January 1973, the Nixon administration suspended almost all existing federal housing subsidies for the poor and announced its intention to search for more effective programs. The housing allowance experiment, then getting under way in 12 selected cities, took on special importance as part of that search.

As EHAP was being completed in 1981, the incoming Reagan administration established a new President's Commission on Housing in February 1981. Its mandate was to review existing programs and research on housing and recommend directions for U.S. housing policy by April 1982. Once again, housing allowances are a major focus of a presidential commission's recommendations.

THE EXPERIMENTAL HOUSING ALLOWANCE PROGRAM

To elicit empirical evidence about the concept of housing allowances, Congress directed HUD to establish EHAP in Title V, Section 504, of the 1970 Housing and Urban Development Act. It authorized $20 million for the program, called for its completion by the end of fiscal year 1973, limited the program to rental housing, and specified the payment formula. In 1974 Section 504 was amended to reflect the program design described below.[2]

The many issues raised about housing allowances defined three basic sets of questions:

(1) How do families respond to housing allowances?
(2) How are housing markets affected by allowances?
(3) How might a housing allowance program be administered, and what are its administrative costs?

These questions were addressed by three different, though interrelated, experiments—the Demand Experiment, the Supply Experiment, and the Administrative Agency Experiment.

THE DEMAND EXPERIMENT

The Demand Experiment focused on how families respond to housing allowances. The specific questions it addressed were as follows:

- Who participates, and how are participation rates affected by program features such as payment levels?
- Does a housing allowance program cause participants to move to new locations?
- What portion of the allowance payment is used for housing?
- Does the quality of housing improve for participating households?
- What are the major differences in household responses between a payment program constrained by housing requirements and a program of unconstrained allowances, such as welfare payments?
- How do housing allowances compare with other housing programs in terms of participation, housing quality attained, locational choices, and costs?

Answers to these questions depend partly on program rules, so the design systematically varied payment levels, housing requirements, and payment formulas to learn how outcomes would differ. The specific variations tested are described below.[3]

The Housing Allowance Demand Experiment (HADE) was the only one of the three to include a control group. To isolate the effects attributable solely to the housing allowances from changes that would have occurred in the absence of any program, HADE collected data on similar families (controls) that did not receive such payments. The difference between the changes in housing conditions of control families and those receiving housing allowances represents the *induced* effects of the program. The HADE design also included a comparison of the effects on similar families of receiving unconstrained payments (with no housing requirements to meet) and housing allowances.

HADE was conducted over a three-year period in two metropolitan areas: Allegheny County, Pennsylvania, the Pittsburgh Standard Metropolitan Statistical Area (SMSA); and Maricopa County, Arizona, the Phoenix SMSA. These sites were selected from among 31 SMSAs by applying a set of criteria that included vacancy rate and size of housing market area as key selection factors. Given the objectives of the experiment, both sites chosen had to be sufficiently large to prevent the allowances to experimental households from having any significant inflationary effect on market rents. Both sites had populations in excess of 500,000 and vacancy rates of about 6% in 1970. The other factors in the set of criteria reflected differences of interest to policymakers. Pittsburgh is a slowly growing, older city with a large black population; Phoenix is a rapidly growing, newer city with a large Hispanic population.

In each of these housing markets, about 50,000 families were surveyed. Representative samples of families eligible by income were selected to be offered participation in the various groups discussed above. In the two sites, approximately 950 control families were selected, and about 2500 families were enrolled in either constrained or unconstrained payment plans.

Through combinations of payment formulas and housing requirements, 17 experimental plans were tested. Two payment formulas were used. The *housing gap formula* can be expressed as $S = b_1C - b_2Y$, where S was the allowance payment; C was the cost of standard housing established for the program which varied by household size, site, and the experimental parameter b_1; b_2 was the rate at which the allowance is reduced as household income increases (or the implicit tax rate), also varied experimentally; and Y was net family income. Under the *percent of rent formula,* the payment, S, was a percentage of the family's rent, $S = aR$, where R was the rent and "a" was the fraction of the rent paid by the allowance. This formula was included to enable estimation of demand functions that describe the way in which expenditures on housing are related to income and the price of housing and other goods.

In combination with the primary payment formula (housing gap), four approaches to housing requirements were tested. Based on American Public Health Association criteria, a set of minimum housing standards was developed as the principal approach. They include 15 major categories of physical attributes that, when taken together, define a decent, safe, and sanitary housing unit. In addition, an occupancy standard limited occupants to two per acceptable bedroom. A second approach, the unconstrained plan, did not require participants to meet any housing standards. As a possible alternative for the physical standards, two other housing requirements were tested; "Minimum Rent High" and "Minimum Rent Low." In these plans, payments were made, regardless of the physical quality of the dwelling, if a household's rent was at or above a required minimum. (The minimum was 90% of C for Minimum Rent High plans and 70% of C for Minimum Rent Low plans.)

THE SUPPLY EXPERIMENT

The purpose of the Housing Assistance Supply Experiment (HASE) was to determine the effects of housing allowances on housing markets. Among the specific questions this experiment was designed to address were the following:

- When all eligible families are offered the opportunity to receive housing allowances, will landlords, developers, homeowners, mortgage lenders, real estate brokers, and others accommodate the recipients in their attempts to improve their housing conditions? Or, as some predicted, will the price of housing simply increase without a corresponding improvement in housing?
- To what extent will housing allowances stimulate repairs, substantial rehabilitation, or the construction of new units?
- As housing allowance recipients attempt to increase their housing consumption by moving, what neighborhoods will they seek and which ones will they succeed in entering? What is the impact on the neighborhoods they leave and enter? Will families move from the central city to the suburbs?

While some of the specific questions of HASE were similar to those of HADE, a different approach was required to answer these questions (see Lowry, 1973, for more details). In HASE, enrollment was open to all eligible families in two metropolitan housing markets. In addition, payments to households were promised for a period of 10 years, as long as the eligibility requirements were met. This long-term commitment was considered necessary to enable demand subsidies to induce supply response.

Both homeowners and renters were eligible to participate in the program. Only one housing allowance plan, a housing gap payment schedule with a minimum standards requirement, was used. The payment formula and standards were similar to those in the Demand Experiment (the standards were modified according to local housing codes of the supply experiment sites).

The two housing markets for HASE were Brown County, Wisconsin, whose central city is Green Bay; and St. Joseph County, Indiana, whose central city is South Bend. These locations were selected from all SMSAs. The housing markets chosen had to differ in certain ways, yet be typical of a substantial portion of markets throughout the nation. Brown County has a rapidly growing urban center, a relatively tight housing market, a good housing stock, and a very small minority population. St. Joseph County has a declining central city with a deteriorating housing stock, a minority population of average size but growing, and an excess supply of central-city housing.

THE ADMINISTRATIVE AGENCY EXPERIMENT

The objective of the Administrative Agency Experiment (AAE) was to gather information about the administration and costs of delivering housing allowances. The design and analysis, therefore, focused on those administrative functions considered of major importance for the administration, management, and operation of a national housing allowance program.[4]

The experiment consisted of experimental housing allowance programs administered by several different types of public agencies and an independent evaluation. The research firm and all agencies performed their roles under contract to and in consultation with HUD. The design provided for natural rather than systematic variations in administrative procedures. Agencies were encouraged to develop their own procedures for administering the allowance program within a set of broad guidelines developed by HUD.

The AAE was implemented as if it were a typical housing program. Eight agencies, which potentially could administer a national housing allowance program, were selected to cperate the experiment. They consisted of two public housing authorities, two county agencies, two state community development agencies, and two welfare agencies. The eight agencies were the Salem (Oregon) Housing Authority; the Tulsa (Oklahoma) Housing Authority; the San Bernardino (California) County Board of Supervisors; the Jacksonville (Florida) Department of Housing and Urban Development; the Commonwealth of Massachusetts Department of Community Affairs; the State of Illinois Department of Local Government Affairs; the Social Services Board of North Dakota; and the Durham County (North Carolina) Department of Social Services.

In the eight agencies, the number of households to which payments could be made was limited to 900 in six cases, to 500 in one case, and to 400 in another. The payment formula was similar to the one used in the supply experiment and to one of the housing gap types in the demand experiment: the difference between the cost of modest standard housing and 25% of the family's preallowance income. Payments were made only to families that either lived in or moved to units that met the program's housing standards.

Figure 1.1 indicates the location of each of the 12 experimental sites and of the two housing allowance demonstration projects that were locally initiated and funded under the Model Cities program.

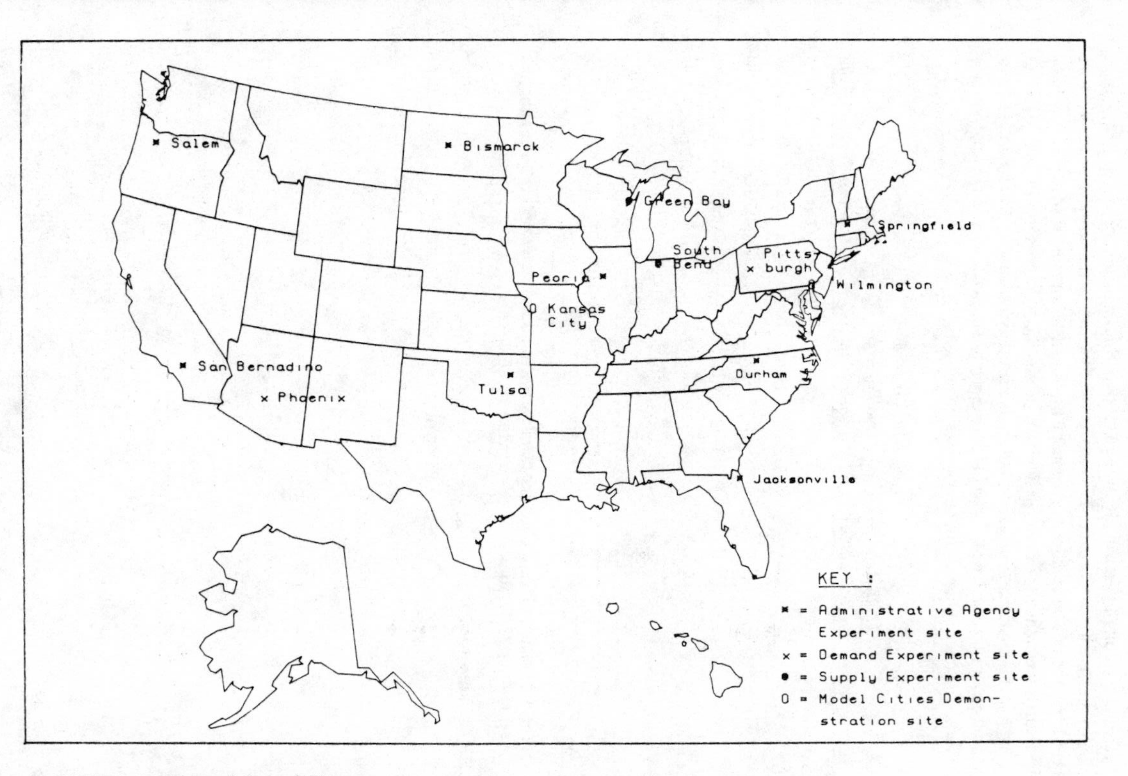

FIGURE 1.1 Experimental Sites

NOTES

1. The Pruitt-Igoe Project, completed in 1956, consisted of 43 buildings on 57 acres near St. Louis's city center. By 1970, the project was so ridden by crime and vandalism and had so deteriorated as a result of the inability of the St. Louis Housing Authority to keep up proper maintenance and repair that it became impossible to find tenants for the many vacant units in the project. At first, the housing authority responded by closing down more than half of the buildings. Eventually the entire project was vacated and demolished (Helbrun, 1981: 369).

2. The initial authorization of $20 million amounted to only one-eighth of the total cost of the experiment. By mid-1980, when most of the research had been completed, the total cost was $158 million. The distribution of this amount was (in millions):

Experiment	*Payments to Households*	*Administration and Program Operation*	*Research and Monitoring*	*Total*
Demand	$ 4	$ 2	$25	$ 31
Supply	40	18	41	99
Administrative agency	10	3	9	22
Integrated analysis	0	0	7	7
Total	$53	$23	$82	$158
Percentage	34	15	51	100

SOURCE: Struyk and Bendick (1981: 297; estimates as of April 1980).

3. See Abt Associates Inc. (1973) for a detailed description of the experimental design.

4. See Hamilton (1979) for a detailed description of the AAE.

2

The Supply Experiment

IRA S. LOWRY

□ THE HOUSING ASSISTANCE SUPPLY EXPERIMENT (HASE) was designed to test the market and community effects of a full-scale allowance program. It was undertaken because many observers doubted the ability of low-income families to bargain effectively in local housing markets and feared that their allowances would be absorbed by rent increases that were not matched by housing improvements, to the detriment of both participants and others. For example, in 1968 the President's Committee on Urban Housing warned:

> The immediate adoption of a massive housing allowance system would be likely to inflate the cost of existing housing considerably, at least in the short run. The large infusion of new purchasing power would result in a bidding up of housing prices for the existing standard inventory. Consequently, any large-scale allowance system would have to be introduced gradually. Such a system might also require strong programs of consumer education and vigorous attacks on racial discrimination to work effectively [pp. 71-72].

Others were interested in or concerned about the neighborhood effects of such a program. Optimists foresaw rejuvenation of deteriorating neighborhoods as landlords repaired rental dwellings to attract allowance recipients and low-income homeowners obtained allowances that enabled them to improve their homes. Pessimists foresaw a general exodus of allowance recipients from deteriorated neighborhoods, leading to the collapse of property values there, and social tensions arising in better neighborhoods as allowance recipients sought housing there. How such a program would affect racial segregation was not clear to anyone; although it would provide minority

---

participants with the means to pay for better housing and neighborhoods, it would not directly address exclusionary practices.

EXPERIMENTAL DESIGN

To explore these issues, HUD authorized ten-year allowance programs in two metropolitan areas: Brown County, Wisconsin (metropolitan Green Bay) and St. Joseph County, Indiana (metropolitan South Bend). The Rand Corporation supervised program operations for the first five years and monitored concurrent events in local housing markets by means of annual field surveys addressed to the owners and occupants of marketwide samples of residential properties. Whereas the Demand Experiment observed the effects on individual households of the "treatment" each received, the Supply Experiment's subjects were housing markets and the treatment was an allowance program.

The experimental strategy in HASE was to apply the same treatment to markets that differed in ways likely to affect outcomes. Because a full-scale allowance program would be expensive, only two sites were chosen—both small in size but contrasting as to market structure and initial condition.

Brown County (48,000 households) had a relatively new housing inventory, its population was growing, property values were high, and vacancy rates were low despite steady growth of the inventory. Less than 2% of the county's inhabitants belonged to racial minorities, so its housing market was unsegregated. St. Joseph County (76,000 households) had an older inventory, including much deteriorated housing in central South Bend; South Bend's population was decreasing, property values were low, and vacancy rates were high. Nineteen percent of the city's households were black or Latin, nearly all of them living in deteriorated neighborhoods where rental vacancy rates exceeded 12% and single-family houses often sold for under $10,000. In each site, we initially estimated that about a fifth of all households would be eligible for assistance.[1]

Identical allowance programs, open to nearly all low-income renters and owners, were operated in the two sites. By the end of the first three program years, enrollment had reached 3600 households in Brown County and 6500 in St. Joseph County. Thereafter, program growth was slow; about a third of all enrollees dropped out each year (usually because they became ineligible) and were replaced by others who were newly eligible. During this steady state, about 8% of all households (about 15% of all renter households) in each county were enrolled in the program.

The allowance programs offered eligible households monthly cash payments calculated on the "housing gap" principle. That is, a household's entitlement equaled the estimated local cost of adequate housing, less one-fourth of the household's adjusted gross income. In practice, this formula worked out to an average annual payment (in 1977) of about $1000 to renters whose gross incomes averaged $4000; and $800 to owners whose gross incomes averaged $4900. At that time, a well-maintained four-room dwelling rented for about $2150 annually, including utilities.[2]

Although enrollment was open to anyone whose allowance entitlement would exceed $120 annually, payments were made only to enrollees whose dwellings met detailed standards as to living space, domestic facilities, safety, and sanitation. These standards were enforced by initial and annual on-site inspections. An enrollee whose dwelling failed could either arrange for repairs or move to an adequate dwelling, and thereby qualify for payments. Recipients whose dwellings failed annual inspections or who moved to inadequate housing faced suspension of payments unless the housing defects were remedied.

The allowance program in each site was administered by a nonprofit corporation created for that purpose, called a housing allowance office (HAO).[3] The two HAOs publicized the program, inviting applications from all who thought they might be eligible. Applicants who passed the eligibility tests were enrolled and informed of their entitlements and of the housing requirements they had to meet in order to receive payments. Participants found their housing on the private market without help from the HAOs and could move or change tenure (renting or owning) without losing their allowances, provided always that their dwellings met program standards.

The HAOs did not set either minimum or maximum housing expenses for participants, and the amount of the allowance did not vary with actual expenses. Participants were entirely responsible for negotiating rents or home purchases, for arranging repairs, and for meeting their financial obligations. The HAOs had no dealings with or obligations to landlords, lenders, repair contractors, or others who might be involved in a participant's housing transactions.

EXPERIMENTAL FINDINGS[4]

During the first five years of program operations, a total of 25,000 households enrolled in the two sites and 20,000 received one or more payments. Four annual surveys of residential properties in each site produced nearly 8,400 records of interviews with landlords, 18,200 records of interviews with renters and homeowners, and 11,500 records of observations on residential buildings. Joint analysis of program and survey data yields the following findings:

- In the mature program, about one-third of those who were currently eligible were currently receiving payments. The main reasons for nonparticipation were the small entitlements of those who were only marginally or briefly eligible and the unwillingness of some whose dwellings were unacceptable to either repair them or move to better housing. The neediest were most likely to participate, but more of them would have participated in the absence of minimum housing standards. However, the standards did prompt considerable housing improvement, as noted below.

- About half of those who enrolled were then living in dwellings that did not meet the program's quality standards. Among those who had to repair or move in order to qualify for payments, about two-thirds did so and one-third dropped out. Overall, 80% of the enrollees eventually qualified for payments. Most of those who dropped out could have recovered repair costs from their first few allowance payments.
- Participation in the program increased the likelihood of occupying standard housing from about 50% to about 80%, and reduced preenrollment housing expense burdens from about 50% of gross income to about 30%. In addition to making required repairs, three-fourths of the owners voluntarily improved their dwellings each year and two-fifths of the renters moved to larger or better dwellings. However, the average participant increased his housing expenditures by only 8% over his estimated expenditures absent the program.
- Enrollees were able to meet program standards without much increase in expenditure because their housing defects were mostly minor health and safety hazards, rather than major structural defects or lack of basic domestic equipment. Repairs were generally made by the participants themselves, their friends, or their landlords, rather than by professional contractors. The average cost of repairing a failed dwelling was about $100, including an imputed wage for unpaid labor. Although allowances augmented the typical renter's income by about a fourth and the typical owner's income by a sixth, they chose to spend only a fifth of the extra money on housing. Thus, four-fifths of all allowance payments were allocated to nonhousing consumption.
- A full-scale open-enrollment allowance program had no perceptible effect on rents or property values in either a tight housing market (Green Bay) or a loose market (South Bend). One reason was that the program increased aggregate housing demand by less than 2%. Another was that it proved relatively easy and inexpensive to transform substandard to standard dwellings. When a renter joined the program without moving, his rent typically increased by less than 2%, even though his landlord may have made minor repairs to bring the dwelling up to program standards.
- The program had little effect on the physical appearance or social composition of residential neighborhoods. Even in neighborhoods where participants made up a fifth or more of all residents, the housing improvements were inconspicuous because program standards were not concerned with cosmetics. Though many renters moved, the origins and destinations of the moves were too diffuse to alter neighborhood populations. The degree of racial segregation did not change perceptibly because of the program.

- After three years of experience with the program, a majority of all household heads and 90% of all participants thought it was a "good idea." Landlords were less enthusiastic, but a majority of those whose tenants included recipients approved of the program. In general, the public approved of who got help, what the help was for, and how the program was run.
- The allowance programs in Green Bay and South Bend were administered by nonprofit corporations under the supervision of Rand and HUD. Hiring staff locally at prevailing wages, these housing allowance offices performed their functions promptly, equitably, and humanely at the surprisingly low cost of $163 per recipient-year. Many of the program's administrative features that contributed to this outcome are transferable to other federal programs.

Reflecting on the experimental evidence, and consulting available national data, we offer the following judgments about the effects of a national program that followed the same design as the experimental one.

(1) Some poor households live in inexpensive and inadequate dwellings; others are adequately housed by dint of spending half or more of their incomes for housing. Housing allowances are flexible enough to remedy whichever circumstances apply to a particular case and can serve homeowners as easily as renters. Nationally, as well as in the experimental sites, budgetary relief is probably a higher priority for low-income households than is better housing.

(2) The public cost per assisted household would be far below that entailed in programs that build new housing for the poor; moreover, we estimate that 85 cents of each program dollar would directly benefit participants. A comparable estimate for the Section 8 Existing Housing program is 57 cents; for the federal public housing program, 34 cents; and for an income maintenance program with no housing requirements, 89 cents.

(3) At most, 10% of all households (half of those eligible) would participate in a permanent national program, at an average public cost of about $1100 per recipient year (1976 dollars), including administration. About 30% of the participants would occupy safer and more sanitary dwellings than they otherwise would, and all would be able to spend more for nonhousing consumption.

(4) We judge that a national housing allowance program would affect only participants and their housing; the broader community would be unaffected for good or ill. Specifically, we think that a program open to all low-income renters is not at all likely to cause significant rent increases for either participants or others, even in moderately tight housing markets. On the other hand, we do not think that a full-scale program would much alter the appearance or social composition of low-income neighborhoods; nor would it much expedite the residential integration of racial minorities.

METHODOLOGICAL ISSUES

MARKET EFFECTS

HUD's Experimental Housing Allowance Program was planned in a period of great ferment in federal housing policy. Those plans were further shaped by a new sense of the possibilities of applying experimental science to the problems of government. Until the mid-1960s, experimentation in government essentially meant launching new national programs whose designs at best reflected theoretical analyses of probable effects but which sometimes provided for systematic evaluation after the programs were operating. The new idea was that the essential features of a contemplated national program could be tested by a carefully designed experiment conducted on a relatively small scale, the results of which would allow much more precise estimation of the effects of the full-scale counterpart and provide valuable guidance on program rules, administrative requirements, and costs if a national program were adopted.

The intellectual model for these social experiments, as they came to be called, was the clinical trial in medical research. In clinical trials, a therapy of unknown effectiveness is administered under controlled conditions to a carefully chosen sample of ailing persons; similar samples get alternative or no treatment. Even if detailed causal links between the treatment and the subject's response cannot be identified, clinical trials enable experimenters to assess the statistical effectiveness of the treatment as against alternative or no treatment of the same ailment.

The Demand Experiment followed this model quite closely. Screened samples of low-income households in two metropolitan areas were offered housing allowances on various terms and conditions. Their responses to those offers and the housing conditions, actions, and expenditures of those who accepted were analyzed as functions of the terms of the offers; and their experiences were compared to those of a control group of households in each site. Thanks to the attentiveness and imagination of a highly capable research team, the comparisons among treatment groups and between treatment and control groups forestalled many attractive but erroneous inferences. Response parameters were estimated with as much precision as sample sizes would support.

As guides to housing policy, these experimental findings were limited in several respects. First, because those treated comprised only small fractions of those in each site who would be eligible for a similar national program, their housing actions rarely impinged on each other and would not be generally noticed as market signals. Second, the housing market context of their actions was neither controlled nor deeply investigated; the analysts could only speculate about reasons for intersite differences in responses. Third, the recruitment of participants and subsequent transactions with them differed substantially from the modes likely to be employed in a national program. Fourth, the maximum term of participation was three years, of which only the first two were analyzed (to avoid misleading termination effects).

These limitations were foreseen and motivated HUD to commission the complementary Administrative Agency and Supply Experiments. The latter's mission was to estimate the market and community effects of a permanent full-scale program. After exploring various alternatives (computer simulation using nonexperimental data, analysis of naturally occurring analogues to the market stimulus expected from an allowance program, microexperiments to test the responses of individual landlords and homeowners to a hypothetical program, and experiments at the neighborhood scale within larger housing markets), Rand and HUD agreed that the best way to learn about the effects of a full-scale program was to conduct one. None of the alternatives seemed likely to provide reliable and generally credible evidence about the effects of an actual program when so little was known about how eligibles would respond to the allowance offer, how they would communicate their housing demands in the open market, and how the suppliers of housing would respond to the resulting market signals.[5]

There were two obvious drawbacks to full-scale experimentation: its expense and the risks it entailed for the host communities. Considerations of cost led us to limit the experiment to two small metropolitan housing markets; risks were managed in two ways—by securing the informed consent of the host communities and by preparing contingency plans that included aborting the experiment after it was under way, if necessary to forestall or limit damage to the community.[6]

Methodologically, the limit on the number of sites was particularly vexing. In the Demand Experiment, the experimental treatment was an allowance offer to an individual household, whose behavioral responses would then be observed. In the Supply Experiment, the treatment was a housing allowance program "offered" to a community, whose aggregate (market) responses would then be observed. From the perspective of statistical inference, the Supply Experiment had a sample size of two. Furthermore, a little thought persuaded us that there was no practical way to identify an appropriate "control group" of sites that could be observed without treatment. However closely we matched the experimental sites with control sites, unpredictable events during the course of the experiment (a flood, the closing of a major industrial plant, or a municipal fiscal crisis) might invalidate the comparison. Comparably monitoring housing markets in a large control group of sites would be impossibly expensive.

In short, the model of the clinical trial was inappropriate for the Supply Experiment. Instead, we chose sites that differed sharply as to market characteristics likely to affect outcomes, and conducted identical allowance programs in each. From administrative records of the program, we could precisely measure the experimental stimulus (number of participants and their allowances) to local housing markets. By annually surveying the markets themselves, we could measure market outcomes (price and quantity of housing services consumed). However, we were dependent on analytical modeling rather than probability theory to distinguish the role of the measured stimulus in producing the measured outcomes, given uncontrollable nonprogram events in each site that could also affect housing markets.

From other essays in this volume and from the final reports of the Supply Experiment, the reader can judge how well we succeeded, as to both measurement and causal attribution. I judge that we established beyond controversy that the market stimulus inherent in a program of this type is much smaller than most observers expected, and for surprising and very important reasons. First, many who are eligible will not participate; second, most low-income families live in dwellings that can be easily and cheaply improved to meet program standards; and third, augmenting low incomes causes only a small increase in voluntary housing expenditures (i.e., beyond the expenditures needed to meet program standards). Further, I can think of no politically plausible variant of the allowance concept that would be likely to generate a substantially larger stimulus to local housing markets.

The measurement of market outcomes as regards housing prices was clouded by the instability of the unit of measurement. The experiment was conducted during a period of rapid national price inflation, led by escalating energy prices that especially affected housing. We established that rents and property values in the experimental sites approximately tracked regional and national indexes, and that the net operating return from rental properties was stable or diminishing during the period of rapid program growth (implying no short-run profits due to increased demand). Multivariate analysis of rent changes for individual dwellings indicates that participants paid a small premium when they brought a dwelling into the program; but if there was any spillover effect on nonparticipants' housing, it was too small to be detected in an inflationary environment (Rydell et al., 1982; see also Chapter 11).

As to neighborhood effects, program records enabled us to measure the direct effects—moves by participants, repairs to their dwellings—with precision, and our field surveys showed us that indexes of neighborhood quality, rents, and property values did not change in patterns that reflected the neighborhood concentration of enrollees or allowance payments. More important, it was clear that the direct program effects, when set against neighborhood aggregates, were too small to perturb neighborhood averages; only a large multiplier effect could have produced substantial neighborhood change (Hillestad and McDowell, 1982; see also Chapter 14).

In short, whatever its benefits to participants, the experimental allowance program did not measurably disturb the housing markets of Brown and St.

Joseph counties during the years of rapid program growth when market disturbances were most likely. That finding was important, because HUD-sponsored computer simulations of allowance programs, one using Brown and St. Joseph counties as examples, came to contrary conclusions (Barnett and Lowry, 1979).

However, the experiment was undertaken to provide guidance for a national program, and a sample of two small metropolitan areas does not provide the basis for statistical inference to other places.[7] Although statistical inference has many advantages as a mode of generalization, it is not the only valid form of inference. Understanding the logical structure of a process enables us to estimate how it would behave in contexts other than the experimental one. We did not, for example, have to launch a thousand rockets to the moon in order to program a trajectory that would reach the target.

As explained above, we observed the joint effects of the allowance program and other events on the housing markets of Brown and St. Joseph counties and found little evidence of program-induced market disturbance. We also explicitly modeled the effects of the allowance program in each site, abstracting from background price inflation and local population and income changes that might have affected actual market outcomes independently of the program. The structure of the model was suggested by our observations on program and market processes; some of its parameters were estimated from HASE data and some from national data gathered in the Annual Housing Survey. Initial conditions for the modeling exercise were those observed in our sites at baseline, and the program's market stimulus was given by the actual histories of participation in each site. The market effects of the program, as estimated by the model, are reasonably consistent with observed market outcomes, once allowance is made for background inflation (Rydell et al., 1982).

To help HUD with the generalization problem,[8] we devised a variant of this model that could subsist on the population and housing market data available from the Annual Housing Survey. HUD compiled the data and ran the model for a hypothetical allowance program conducted in a national sample of 20 metropolitan areas, varying the key parameters around the values estimated for the Supply Experiment. The results indicated that only in exceptional circumstances would an allowance program significantly perturb a local housing market.[9]

OTHER RESEARCH TOPICS

After the Supply Experiment was under way, it became apparent that its design offered opportunities for complementary research on issues originally assigned to the Demand and Administrative Agency experiments: the determinants of participation, effects on participants, and administrative effectiveness and efficiency. These studies were added to our agenda with only minor modifications of the data collection plan.

The Supply Experiment offered an unusual opportunity for participation analysis, in that our marketwide household surveys obtained enough data from respondents to determine their eligibility under program rules. Within the limits of sampling variability, we therefore had a solid base for the measurement of participation rates. Whereas the Demand Experiment individually invited a screened sample of eligible households to enroll, the Supply Experiment extended a general invitation (without time limit) to the public—a mode close to that of a permanent national program. Thus, one might expect the participation experience in the two experiments to differ. Because enrollment was open throughout the experiment, we were able to observe turnover (not just attrition). For all these reasons, HASE participation studies (Ellickson, 1981; Carter and Balch, 1981; Coleman, 1982; Wendt, 1982; Carter and Wendt, 1982) provide a valuable complement to those of the Demand Experiment, which focus on the effects of program variation.

With respect to effects on participants, the main advantage offered by the Supply Experiment was large samples. In the two sites combined, over 25,000 households enrolled and, of these, over 20,000 qualified for payments. We had complete dossiers on each case from the time of initial application. However, there were also disadvantages; an open enrollment program does not allow for a control group—households similar to the experimental subjects but not permitted to participate. Furthermore, there was only one allowance program design for the Supply Experiment; the effects of program variations could not be tested.

The control problem was handled analytically, by a Latin-square design (Mulford et al., 1982). From our baseline (preprogram) household surveys, we retrospectively identified households which later became allowance recipients. We compared their preprogram housing consumption with that of other baseline households, controlling on income and demographic characteristics. We found that, except for a small intercept shift, future recipients' consumption responded to the same factors in the same way as did the consumption of those who never joined the program. Taking account of that intercept shift, as well as of temporal shifts in other parameters, we were able to estimate how much housing all allowance recipients would have consumed absent the program. This analytical control method is less foolproof than using a contemporaneous control group as a benchmark, but it guards against the most prominent dangers of before-and-after comparisons.

The administrative studies (Kingsley and Schlegel, 1982; Kingsley et al., 1982; Tebbets, 1979; and Rizor, 1982) were based on detailed time-and-task records maintained by the HAOs under Rand's supervision, and on quality control programs for eligibility and housing certification. The former enabled us to estimate with unusual precision the cost per case of each step in administrative processing, and thereby to highlight opportunities for administrative change or program redesign that would save money without reducing program effectiveness; however, such variations were not systematically tried. The latter enabled us to estimate the incidence of errors in allowance entitlements and housing inspections and their fiscal and other consequences.

INCIDENTAL BENEFITS OF THE EXPERIMENT

In order to estimate program effects on the housing markets of our two sites, we compiled detailed time-series on key market variables and analyzed market structure and processes. Our observations led to some new insights into market processes and parameters that are both theoretically and practically important.

The annual surveys of residential properties conducted in Brown and St. Joseph counties were designed to measure changes in the characteristics of the housing inventory, its utilization, the cost of supplying housing services, and the prices charged for them. Our sample design yielded annual marketwide probability samples of households, dwellings, properties, and landlords; and time-series on specific properties, including their current owners and occupants. For each property in the sample, we compiled detailed annual accounts of both operating and capital expenses; for rental properties, the data cover both tenant and landlord outlays, as well as accounting for rental revenues and vacancy and collection losses (Neels, 1982a, 1982c). We know of no other marketwide survey that provides comparable financial detail.

From these data, we were able to estimate hedonic indexes for housing attributes (Barnett, 1979; Noland, 1980); the income elasticities of housing expenditures for both renters and owners (Mulford, 1979) and the elasticities of demand for specific housing attributes (Barnett and Noland, 1981); a four-factor (land, improvements, energy, building services) production function for housing services, including the elasticities of substitution between the factors (Neels, 1982b); the price elasticity of the rental occupancy rate (Rydell, 1982); and the determinants of housing repair and improvement policies (Helbers and McDowell, 1981).

The parameters we estimated are, of course, specific to our sites, and their generality remains to be confirmed by replication elsewhere. But the insights they suggest are powerful ones. Without going into detail, a few examples are in order:

- The cross-sectional income elasticity of housing demand is far below the long-run aggregate elasticity. In other words, cross-sectional variation of income about the mean has less effect on housing consumption than a long-run change in the average income of all households.

- The existing inventory of housing is flexible in response to demand changes. The output of housing services, as valued by the market, can be substantially increased or decreased by varying current inputs (energy, building services, repairs) without great loss of efficiency.
- In rental housing, rents vary surprisingly little with market condition. Imbalances between supply and demand tend to be reflected in vacancy rates rather than remedied by price changes. Property values, however, are quite sensitive to rental revenue, which reflects both price and vacancies.
- With respect to the production and consumption of housing services, submarkets are not salient; the flow prices of housing attributes are about the same throughout the market. Investment submarkets, however, are quite distinct; physically comparable properties in different neighborhoods may differ in market value by a factor of two for a long time.

DISAPPOINTMENTS

As one of the few who participated in the Supply Experiment from the beginning to end, I formed expectations about what might be accomplished, some of which were disappointed. Other participants and observers, with different expectations, doubtless were differently disappointed, but it seems worthwhile to call attention to a few aspects of our research whose yield seems to me less than it might have been.

My principal disappointment is that we were not able to exploit all the opportunities for useful research that were offered by the HASE data files. That outcome is less attributable to the topical limitations of our charter than to the time and expense entailed in converting raw data to clean, well-organized, well-documented research files. The HASE analysts "practiced" on early data from the allowance programs and the field surveys, and what they learned thereby greatly affected both the subsequent research agenda and the way data files were assembled and managed. But they had less than a year to operate on the full data sets before time and money ran out.

Fortunately, the data are preserved for others. Along with data from the other experiments, the HASE files—8 five-year files of program data and 32 files of survey data—and their documentation were deposited in HUD's Housing Research Data Center, where they will be accessible to the public.[10]

A more specific disappointment was the low yield of our research into residential mobility. How the allowance program affected the mobility of participants and the composition of neighborhood populations was a topic included in our initial research charter. Although our household surveys did not follow movers (the sample element was a dwelling whose current occupants were interviewed), we did obtain a five-year mobility retrospective on each household that entered the sample. Although the HAOs did not directly record moves as dated events, the approximate dates of moves by participants could be inferred from their housing evaluation records. Despite a large amount of data on movers and their circumstances before and after moving, we never developed a powerful model of residential mobility as an economic or social process.[11]

We were much more successful at modeling market adjustments to shifts in demand. As noted above, the data led us to some analytically powerful insights exploited in a series of theoretical and empirical papers, mostly by Rydell (1979a, 1979b, 1980, 1982). However, we never achieved micro-models of consumer and producer behavior that rigorously supported our macro-model, the parameters of which we estimated from HASE and AHS data. There were times when micro-macro integration seemed in reach; but each possibility faded under close scrutiny.

Finally, we contributed little to the theory of tenure choice. Nearly all our work follows the tradition of treating renters and homeowners as though they were different species. For me, this disappointment is mitigated by the observation that our data collection effort was not designed to serve the analysis of tenure choice; and further, that the separate species assumption was adequate for program analysis. But I had hoped for a wider model of consumer behavior than we achieved.

VALEDICTORY

Of the "social experiments" undertaken by the federal government in the 1960s and 1970s, the Supply Experiment was the largest in number of participants, longest in duration, most expensive, and operationally most complex. Nearly everyone who has paid attention to it believes important things were learned from it, but not everyone agrees that, on balance, it was worth the trouble and expense. That judgment surely should depend on the scientific and political consequences of the experiment, which cannot be fully evident at the moment of its completion, and may never be clearly attributable to it.

This book is one among several vehicles for placing the methods and findings of the EHAP experiments before the public for assimilation and application. As one who was present at the creation, I await the outcome with interest.

NOTES

1. The most detailed description of the two sites at baseline is given in Rand Corporation (1977: Sec. IV).

2. These cross-site averages conceal significant differences between Brown and St. Joseph counties. In the latter, incomes were lower and payments were higher. Rents for comparable dwellings were about the same in the two counties, though property values were much lower in St. Joseph County.

3. The administrative regulations governing the program were developed jointly by Rand

and HAO staffs and are documented in the *Housing Allowance Office Handbook* (Katagiri and Kingsley, 1980).

4. The following section is taken from the executive summary of the HASE final report (Lowry, 1982a). The findings are detailed in the full report (Lowry, 1983).

5. The history of the HASE experimental design can be traced through a series of reports first published in 1971-1973 but subsequently republished in the years indicated by the citations: Lowry et al. (1981); Lowry (1980b, 1980c, 1980d); HASE Staff (1980, 1981). The authoritative account of the final design is Lowry (1980b).

6. See Lowry (1980a). As it turned out, the experimental allowance program did not perturb housing markets in either site in ways that bothered local residents, so the contingency plans were never exercised.

7. This limitation, incidentally, applies also to social experiments more closely modeled on clinical trials. Those conducted to date have, for practical reasons, chosen their subjects in only a few places; there is no guarantee that similar subjects would behave identically in different local contexts, especially if the reasons for their observed behavior are not well understood.

8. Generalizing experimental findings to a national program was not part of Rand's charter for the Supply Experiment. Originally, that task was assigned to The Urban Institute.

9. The unpublished analysis was conducted in the spring of 1981 by Howard Hammerman, Office of Policy Development and Research, and is cited here with his permission. The underlying model of short-run market adjustment is presented in Rydell (1980).

10. The files are described in a three-volume *User's Guide to HASE Data* (Hansen et al., 1982).

11. We did, however, produce a useful study of housing search by renters who moved (McCarthy, 1979); it relates search techniques and outcomes to household characteristics.

3

The Demand Experiment

STEPHEN D. KENNEDY

☐ THE FINDINGS of the Demand Experiment are presented below in terms of three broad areas—the evaluation of housing need, comparisons of allowances (or vouchers) with other housing programs, and the impact of allowance programs in terms of participation and housing change—followed by a brief critical discussion of methodology.

HOUSING NEED

Recent analyses of census data have suggested that physically inadequate housing might be disappearing in the United States and that the remaining housing problem is almost entirely a matter of high housing costs and low income (see Weicher, 1976; Congressional Budget Office, 1978). Budding's (1980) analysis of detailed housing information collected in the Demand Experiment makes it clear that physically inadequate and/or overcrowded housing conditions are far more common than the available national data would indicate. In terms of contemporary standards, Budding found that over half of the low-income renters enrolled in the Demand Experiment lived in units that were either physically inadequate or overcrowded. Over two-thirds of the low-income enrollees were carrying high rent burdens (rents in excess of 25% of income). Only 12% of the low-income households escaped both physically inadequate or overcrowded housing and high rent burdens (Budding, 1980: chap. 1).

In evaluating these figures it is important to understand that Budding set out to describe housing conditions in terms of the contemporary standards defined by housing planners and program managers. This has three important consequences. First, despite widespread failure by contemporary standards, U.S. housing has improved dramatically over the last 50 years. Indeed, much of the difference between Budding's results and those of other analysts simply reflects the fact that the housing quality information collected in the Demand Experiment was detailed enough to allow relatively unambiguous application of contemporary standards.

Second, violations of contemporary standards defined by housing planners do not automatically establish a need for specifically housing-oriented programs. Proponents of general income assistance can still argue that poor housing is simply one of the many problems associated with low incomes and that there is no reason to create programs specifically aimed at housing instead of generally increasing incomes. While Budding's analysis does refute the assertion that there is no housing problem, it does not address several aspects of housing and neighborhoods that might justify programs specifically tied to housing, including the general benefits that may accrue from programs aimed at improved neighborhoods, support of the construction industry, or reduced racial and economic segregation or a need to design program benefits to take account of substantial variations in the costs of adequate housing and/or nonfinancial barriers to adequate housing.

Third, while the use of contemporary standards, and in particular the inclusion of rent burden targets in assessing the need for housing assistance, defines a potential population far larger than current funding levels can reach, the extent of need varies considerably within this population. Almost half of the households with incomes below poverty suffered from both poor housing and high rent burdens as compared with about a seventh of the low-income households with incomes at or above poverty. Likewise, most severe physical deficiencies were concentrated among the very poor (Budding, 1980: chap. 4).

Failure to recognize such variations in need leads to unsystematic allocations of limited program funds across households with widely varying degrees of need. Programs of low-income housing assistance typically start by setting a general goal of providing decent housing at reasonable cost. The need for the program and its potential eligible population are then typically defined by households that either live in substandard housing or pay a larger than "reasonable" fraction of their income for rent. But housing programs in the United States are never funded at anywhere near the levels needed to serve their entire eligible population. In Pittsburgh and Phoenix, for example, about 23% of the low-income renter households were in subsidized housing in 1975. Nationally, the U.S. Department of Housing and Urban Development (1973: 4-27ff) has estimated that in the early 1970s all low- and moderate-income programs combined served less than 10% of eligible households at any income level.

There are at least three strategies for determining which households receive assistance. One that may most closely resemble current practice is simply to raffle off places on a first-come, first-served basis. This has a certain crude equity, since all households get an equal chance to receive benefits. Given Budding's analysis, however, it also means that assistance will be given to some households in moderately difficult straits and denied to others in much worse housing.

A second strategy is to target assistance to demographic groups in greatest need. A program targeted at all low-income renters with physically inadequate or crowded housing (as defined by Budding) and rent burdens greater than 40% of income would potentially encompass about 12% of the low-income renter population in Pittsburgh and Phoenix. This is less than the proportion currently living in subsidized rental housing in these cities. The difference would be in who is served. Under a poor housing/severe rent burden criterion, almost all (90%) of the target population would be households in poverty. Under current programs, less than half (49%) of subsidized housing tenants in Pittsburgh and Phoenix were in poverty.

A third strategy, one not yet employed, is to offer programs that are consistent with the funding levels provided. A public housing program that offered only low- or moderate-quality units while charging rents equal to 40% to 50% of tenant income would appeal only to households in very poor housing or with very high rent burdens or both. The advantage of such a program is obvious. If tenant contributions are made high enough and unit quality held low enough, the program can serve all households that want its assistance and will appeal only to households in the greatest need. This would both direct assistance to those households in greatest need and concentrate program effects on replacing the least adequate housing. The disadvantage is equally obvious. It is difficult to announce publicly the implications of low program funding levels that leave people in clearly unacceptable housing at outrageous costs.

ALTERNATIVE PROGRAMS

Mayo et al. (1980a, 1980b) present an extensive cross-sectional analysis of the housing provided by, and costs associated with, a wide range of housing programs. These include both the housing allowance programs tested in the Demand Experiment and the other major low-income rental housing programs in existence at the time of the Demand Experiment (public housing, Section 236, and Section 23 Existing Leased housing). Indeed, in total, the Demand Experiment studied over 40 different program options. These can be grouped into nine major alternatives:

- no program (represented by unsubsidized control households);
- expanded welfare or income maintenance programs (represented by the unconstrained plan);
- rent rebates (a form of "housing stamp" program represented by the percent of rent plans);
- two forms of minimum rent housing allowances (High and Low plans);
- a minimum standards housing allowance;
- the Section 23 existing leased housing program (including both the original and revised Section 23 program);

–two construction programs—public housing (including both conventional and turnkey programs and elderly and family projects), and Section 236 (including participants with and without rent supplements, elderly and family projects, limited dividend and nonprofit sponsors, and newly constructed and rehabilitated projects).

The major differences among these programs lie in the extent to which they tie assistance to specific types of units and in the mechanisms used to obtain housing. No housing requirements were imposed under the unconstrained or percent of rent allowance programs. Minimum rent allowance programs set a floor on participant housing expenditures but had no requirements for physical or areal adequacy. The other programs—minimum standards housing allowances, Section 23, public housing, and Section 236—involved explicit physical unit and neighborhood requirements, though the exact standards used varied from program to program. Under the various allowance programs, participants arranged for their own housing, using the private rental market. Under Section 23 (existing), housing was leased in the private market by the local housing authority and then sublet to participants, though there was also a revised Section 23 program in Phoenix which operated more like housing allowances. Under public housing and Section 236, housing was constructed for the program and then rented to participants.

On the other hand, most of the programs essentially offered their participants similar levels of financial assistance either in the form of direct cash payments or reduced rents. The major exception was Section 236, which offered much lower benefits than the other programs unless combined with assistance under the rent supplement program.

These differences in payment schedules, housing requirements, and program mechanisms are reflected in differences in participant housing and program costs. All of the programs examined appear to offer participants relatively similar overall levels of housing. The estimated average private market rental value of units under the eight program groups (based on hedonic indices; see Merrill, 1980) were always within 10% of the value for minimum standards participants (Kennedy, 1980: 66). Despite the overall similarity in rental values, however, program outcomes varied substantially in terms of physical standards, crowding, rent burden, location, tenant satisfaction, and costs.

Differences in physical adequacy and participant rent burdens largely reflect differences in program rules. In terms of Budding's measures of physical adequacy, for example, participants in unrestricted cash transfer programs generally occupied housing similar to that occupied by unsubsidized control households. The major difference in the housing situation of participants in unrestricted programs was the lower rent burdens that resulted from having levels of housing expenditures similar to those of unsubsidized control households, offset by fairly substantial monthly subsidy payments. The same pattern was observed for participants in general rental assistance programs without explicit physical housing requirements, such as the percent of rent or minimum rent programs. The only exception to this was the minimum rent high allowance program in Phoenix, which generally resulted in somewhat higher quality levels than those found among unsubsidized households.

Even among programs with explicit physical housing requirements, there were sometimes substantial differences in the physical adequacy of units, depending on the relationship between the standards used to assess the units and the requirements imposed by the program. In general, programs ranked highest when they were evaluated by their own requirements, and as evaluation standards deviated from program requirements, rankings would frequently reverse. Thus, for example, the HUD minimum property standards imposed on public housing and Section 236 were generally more stringent than the minimum standards required in the Demand Experiment. Nevertheless, a larger proportion of minimum standards than public housing or Section 236 recipients lived in housing that passed standards similar to minimum standards. On the other hand, more public housing and Section 236 units passed other standards. These reversals in program rankings were much more dramatic in Pittsburgh than in Phoenix, but the pattern held in both sites.

Differences in rent burdens among the different programs were also directly related to program rules. Minimum rent allowance programs, which did not allow households with unusually low rents to participate, generally had participants with higher rent burdens than other programs. Even higher rent burdens were encountered in Section 236. Because of this, Section 236 assistance frequently included rent supplement payments, which tended to bring rent burdens into line with those found in other assistance programs.

While explicit imposition of housing standards or rent burden rules does result in important differences in the housing situation of participants, the differences appear to be closely tied to the explicit requirements used in the program. This suggests that if programs are to be assessed in terms of specific housing standards, the standards used should be justified as directly desirable and not as proxies for more general notions of adequacy. Given the similar program rental values, it also suggests that any reasonable standard can be met by imposing it explicitly on the program, regardless of whether units are obtained by households or public agencies, from existing units, or by new construction.[1]

Differences with respect to unit location and tenant satisfaction, on the other hand, seemed to be more deeply embedded in program mechanisms. With the exception of Section 236 in Phoenix, units in the programs with locations selected by the government (Section 23, public housing, and Section 236) on average were located in lower income neighborhoods with higher minority concentrations than those selected by housing allowance recipients. Only public housing, however, shows strong evidence of actually moving participants into areas with higher concentrations of minorities or low-income households than those they would normally have occupied. In other programs, and to some extent in public housing, much of the effect of restricted locations seems to be on who participates rather than where they live. Analysis by Mansfield (Mayo et al., 1980a: chap. 5) indicates that participants in all programs tend to come from similar areas to those offered by the program rather than being forced into more heavily minority or low-income neighborhoods.

Analysis by Warner (Mayo et al., 1980a: chap. 6) found that taking account of participants' characteristics and housing situations, there were no significant differences in expressed satisfaction between minimum standards housing allowance participants and households in elderly public housing or Section 236 projects or for Section 23 households in general. There were substantially lower levels of satisfaction for households in public housing and Section 236 family projects.

By far the most dramatic differences among the programs are in their costs. Relative program costs were analyzed by Mayo (Mayo et al., 1980b), whose analysis takes account of both budgeted costs such as payments and operating costs under housing allowances, debt service, maintenance and operating costs, and payments in lieu of local taxes under public housing, and unbudgeted indirect costs such as the loss in federal tax revenues involved in financing public housing with tax-exempt bonds or in the accelerated depreciation provisions associated with Section 236. The results are startling, though generally consistent with the patterns of relative program costs found in other analyses.

Both costs and cost allocations vary substantially. In terms of cost allocations, the major differences are a relatively larger local government contribution (in the form of property tax abatements) for public housing and a relatively larger share of costs borne by tenants under Section 236 (without rent supplements). These cost allocations reflect the specific funding mechanisms adopted for each program and could in theory be adjusted at will. Thus, for example, the addition of rent supplements to Section 236 transfers a substantial share of costs from tenants to the federal government.

More important are differences in total program costs. Each of the programs studied in effect divides the cost of tenant housing among the federal government, local governments, and tenants. These costs will tend to be larger than the simple rental value of recipient housing, if only because they must cover the administrative costs of the program. Thus, for example, Mayo et al.'s analysis suggests that program costs per unit under a minimum standards housing allowance might exceed rental values by from 9% to 15%, almost entirely because of the costs of program administration and nonfinancial services to enrollees. Similar comparisons for new construction programs (public housing and Section 236), however, show costs for newly built

(1975) units ranging from one-and-a-half to more than two times the estimated market rental value of the housing provided. Overall, estimated annual costs required to obtain additional units under construction programs in 1975 were two-thirds again as high as those estimated for minimum standards housing allowances. Given fixed total budgets with identical tenant contributions, construction programs could serve only six families for every ten assisted by housing allowances (Mayo et al., 1980b: chap. 5).

Although the large excess costs encountered in new construction programs were estimated for only two sites (Pittsburgh and Phoenix), they are consistent with other studies involving different cities (Mayo et al., 1980b: 70-74). Furthermore, they hold up under reasonable projections of trends in inflation and depreciation over the life of new construction projects (pp. 75-94). These results clearly confirm the hypothesis that programs such as housing allowances, which make use of the existing housing stock, can provide similar housing at far lower costs than new construction.

Furthermore, while Mayo et al.'s analysis cannot deny the possibility of important construction and operating inefficiencies in public housing and Section 236, they suggest that a major portion of the excess costs in these programs is due to market forces. The relative price of renting existing units as opposed to building new ones has declined considerably over the last 25 years. As a result, the excess costs involved in constructing new units have grown; indeed, Mayo et al. suggest that some construction projects in Phoenix may have yielded reasonable returns as late as 1965.

This hypothesis, if correct, has three important implications. First, it suggests that however valuable they may be, attempts to improve the efficiency of construction programs are not likely to reduce costs enough to overcome the underlying difference in market prices. Second, given a rapid enough inflation in rents relative to construction costs and interest rates, construction programs could provide a cheaper means of providing low-income housing than housing allowances. Indeed, this could hold true even if public construction was less efficient than private construction. Finally, Mayo et al.'s hypothesis of the role of market inefficiencies in construction costs suggests that, in theory, the government might be able to purchase existing units for public housing at no greater (eventual) cost than leasing them. Equally important, if most of the excess costs of construction programs arise at construction, only modest savings would be realized from the sale of already built units and conversion of subsidies to housing allowances.

PROGRAM IMPACTS

Minimum standards housing requirements exclude substandard units from program subsidies. Analysis of program participation by Kennedy and MacMillan (1980) showed that they effectively exclude many of the households initially in substandard units as well. Among households offered enrollment in the minimum standards allowance plans, 38% accepted the enrollment offer and participated in the program. The participation rate for the percent of rent and unconstrained programs, which had no housing requirements, was 84%, over twice as large. Furthermore, most of the households that did participate in the minimum standards allowance programs already lived in minimum standard housing or were about to move to such housing on their own. Among households that would not normally have lived in minimum standard housing without the allowance program, less than 20% participated. Among those that already lived in or were about to move to minimum standard housing, almost 80% participated (Kennedy and MacMillan, 1980: 24-25, 127-132).

As a result, demographic groups that were relatively less likely to live in minimum standard housing were also relatively less likely to participate in a minimum standards allowance program as compared with a percent of rent or unconstrained program. Under a housing gap payment formula, participation among larger and poorer households is increased by the fact that they receive larger allowance payments. Both housing gap and unconstrained programs tend to have greater initial appeal for these groups (as would a rent rebate appropriately conditioned by income and household size). The imposition of housing requirements at least partly undoes this effect—disproportionately reducing participation among minorities, large households, and the very poor.

Participation rates under housing gap allowances can be increased by using less stringent housing requirements or offering higher payments. Thus, for example, less than a third of eligible households lived in minimum standard housing, and only 38% of the households offered enrollment in programs with this requirement participated. In contrast, more than two-thirds lived in housing that passed the minimum rent low requirement, and 60% of households offered enrollment in programs with this requirement participated (Kennedy and MacMillan, 1980: 33-139). Of course, less stringent requirements also mean there is less difference in the housing of housing allowance and unconstrained recipients.

Higher payments can also increase participation. If average payments offered under the minimum standards program had been doubled from about $800 to roughly $1600 per year, the participation rates among households that would not normally have lived in minimum standard housing would have increased from less than 20% to more than 40%, and overall program participation would have been about 56% instead of 38%. Alternatively, payment schedules could be changed to adjust participation among various demographic groups while leaving the overall participation rate unchanged (Kennedy and MacMillan, 1980: 132).[2]

Under a universal entitlement program, in which all eligible households may qualify for payments, participation rates directly determine the number of households reached by the program and the total program costs. Outside of the allowance experiments, however, no housing program in the United States has ever been run on a universal entitlement basis. Thus, the fact that most households would not participate if the program were fully funded may be irrelevant. If program funding levels exclude 90% of eligible households, there seems little reason to worry about the fact that housing requirements exclude 60%.

Even if the level of participation under a housing gap allowance is not of concern, however, the differences in participation among different demographic groups may still be important. These can, of course, be offset by judicious selection of applicants. If under a limited entitlement program relatively few very poor households apply for the program and qualify for payments, the imbalance in recipients might be remedied by selecting more very poor applicants for enrollment. However, applicant selection would be unlikely to correct the imbalance between households that would normally live in housing that meets program requirements and those that would not, even though some demographic groups are less likely to live in required housing than others.

The strongest predictor of whether or not a household is planning to live in minimum standard housing is whether or not it is already in it. Among unsubsidized control households that already lived in minimum standard housing at enrollment, 83% were still in minimum standard housing two years later, as opposed to only 18% of the households that were in substandard housing at enrollment. One strong selection criterion would be to confine program benefits to households that were in substandard housing before they joined the program. But even this extreme selection of applicants would not completely undo the relative advantage of households that would normally occupy minimum standard housing. In the minimum standards allowance programs tested in the Demand Experiment, 66% of recipients were households that would normally have occupied minimum standard housing. Excluding all households that already lived in minimum standard housing when they enrolled would have reduced this proportion, but only to 46% (Kennedy, 1980: 174-175).

It seems likely, therefore, that even on a limited enrollment basis, allowance programs would tend to appeal much more strongly to households that would normally meet housing requirements on their own and would draw many or most of their recipients from among such households. The reasons for such widespread nonparticipation are not clear. It appears that perhaps 20% of households simply are not interested in the transfers offered regardless of whether they are tied to housing (Kennedy and MacMillan, 1980: chap. 3). These households generally would receive small payments; most frequently mentioned reasons given for not participating were that they did not like to take money from the government or that the various reporting requirements were too bothersome. What is more worrisome is the additional 60% of households in substandard housing that did not participate. While this nonparticipation is not well understood, it does not seem to be connected with any reluctance to move per se and may partly reflect difficulties in finding acceptable units at an affordable price (Kennedy and MacMillan, 1980: 125).

If difficulties in search are the cause of substantial nonparticipation, there would seem to be a role for publicly owned housing, which could be used as a known source of units at a fixed (in terms of income) price. Given Mayo et al.'s analysis, the proper course would appear to be either to use the existing stock of publicly owned units (many of which may be reasonably efficient to operate, given that the initial construction costs are sunk) or if necessary to purchase or lease existing private units as they are. Thus, we are driven back toward the original Section 23, if only as a maintained alternative.

The issue of nonparticipation in voucher programs is part of a more general issue of availability. Michael Stegman once suggested to me that in middle-sized southern cities prior to 1965 at the earliest, there often simply was no decent rental housing available to blacks at any price. This situation supposedly no longer prevails or is at least less blatant. There remains, however, clear reason to be concerned with the access of minorities to housing. Briefly, the allowance experiments confirmed the fact that racially segregated housing patterns, for example, are almost totally independent of differences in minority and nonminority income and rent; the provision of money does not undo racial concentration (Atkinson et al., 1980). Analysis of Pittsburgh data by Vidal (1980) found little direct perception of discrimination and no apparent effect of perceived discrimination on the search and/or final locations of minority households. What Vidal did find was that search itself mimicked the pattern of racially concentrated housing. Further, while some of this mimicking appeared to reflect the fact that most households find their units through informal contacts with friends and relatives, Vidal noted that minorities were more dependent on realtors in search than nonminorities.[3]

These results could well be tested in other cities. They suggest potentially important limitations in the access of minority households to housing and ones which are not likely to be remedied by simple provision of support in discrimination cases or housing vouchers.

The patterns of participation found in the Demand Experiment are not only important in determining which households are helped by a program but also have strong implications for the nature and extent of program impacts on housing. A housing gap form of housing allowance essentially divides households into two groups—those that would normally occupy housing that meets the program requirements (or would normally do so given the extra income provided by the allowance) and those that would have to change their normal housing to meet requirements. The first group is not constrained by the housing requirements. Because they would meet the requirements anyway, they are free to use the allowance payment in the same way they would use an unconstrained income transfer. As a result, these households participate in an allowance program at about the same rate at which they would participate in an unconstrained program. Likewise, an allowance program would not be expected to change their housing any differently from a similar unconstrained program.

On the other hand, households that would not normally meet requirements do have to change their housing from normal patterns. Many, and frequently most, of these will not participate. Those that do participate, however, will have to change their housing differently, and supposedly more, than they would under an unconstrained program.

This pattern of relative program impacts is quite intentional. Once households are in acceptable housing, there may be little reason to require them to spend more to occupy even better housing. A housing gap allowance is designed to get households into acceptable housing, or at least to offer them decent housing at affordable cost, while offering them as much discretion as possible about how to live and how to spend their limited resources. Assistance is directly targeted toward housing improvement only for those that need to change their housing in order to obtain what the program defines as decent living conditions. For other households, many of which have obtained decent housing at the cost of high rent burdens, the allowance offers general financial assistance in supporting their housing costs.

This targeting of housing change fails, however, to the extent that the allowance is unable to reach households in substandard housing. The larger the proportion of recipients drawn from households that would normally occupy standard housing, the more the allowance program would be expected to have the same effects as an unconstrained transfer. It is true that all minimum standards recipients occupied minimum standard housing, as compared with only 32% of unconstrained transfer recipients. It is also apparent that unconstrained payments had little or no effect on the proportion of households in minimum standard housing. This does not, however, mean that the allowance program placed an additional 68% of its recipients in minimum standard housing. Two-thirds of minimum standards recipients were households that would normally have lived in minimum standard housing. Thus, the minimum standards requirements moved one-third of its recipients from substandard to minimum standard housing. The difference between the two programs' impacts on the housing of recipients is not, therefore, as great as simple comparisons of recipient housing would indicate.

Moreover, differences between the two programs are much smaller when they are projected to universal entitlement programs. Under universal entitlement, an allowance program would still have the same larger effect for recipients, but fewer recipients. As a result, program effects on the entire population would be much smaller than those for recipients. In comparison to a similar unconstrained program, the minimum standards programs tested in the Demand Experiment, would, for example, have reached less than half as many households and increased the proportion of eligible households in standard housing by about 12 percentage points (from 31% to 43%).

But this is not the end of the story. Analysis by Friedman and Weinberg (1980b) of the housing changes generated by the various allowance programs suggests that minimum standards requirements might be better characterized as generating different rather than more housing changes. Friedman and Weinberg compared the housing of minimum standards recipients with that of unsubsidized control households and unconstrained recipients two years after enrollment in terms of housing expenditures, estimated market value, and the proportion of households in physically adequate housing. They took account of the effects of differential participation, so that their estimates reflect program impacts on recipients rather than simply comparisons of recipient housing.

Like Kennedy and MacMillan, Friedman and Weinberg found that the minimum standards requirements moved about a third more of its recipients from substandard to minimum standard units than the unconstrained program. They also found that minimum standards recipients that met requirements after enrollment had much larger changes in housing expenditures above normal levels than households that were already in minimum standard housing when they enrolled. Nevertheless, the overall increase in expenditures above normal levels was almost identical for minimum standards and unconstrained recipients—about 10% in each case (1980b: 118-122). Furthermore, when housing was evaluated in terms of Budding's measures, the two programs again had very similar impacts on recipient housing.

The percentage of recipients passing either of the two standards based on Budding's classifications was, of course, much larger for the minimum standards programs. However, this difference was almost entirely due to the fact that minimum standards requirements kept many households in substandard housing out of the program. In terms of changing recipient housing, about 15% of the minimum standards recipients were moved from unacceptable to acceptable housing, based on Budding's measures. The estimated impact for unconstrained households was only marginally (and statistically insignificantly) lower (see Friedman and Weinberg, 1980b: 25-27; Kennedy, 1980: 155-186).

It appears, then, that the extra housing change induced by the minimum standards requirements over a similar unconstrained program very specifically focused on the requirements themselves. Indeed, it is possible that, in comparison to the unconstrained program, the minimum standards requirements did not induce additional spending on housing or overall increases in the market value of recipient units, but rather reallocated expenditures to obtain the specific features imposed by the requirements (see also Merrill and Joseph, 1980: 87-92).

The remarkable specificity of requirements effects found by Friedman and Weinberg presents serious problems for the design of a minimum standards housing allowance. While there are no doubt conditions that would be almost universally regarded as unacceptable in the context of modern American standards, most housing standards both include items many people would find unnecessarily burdensome and omit items the same people would regard as critical. Given the extent to which the effect of minimum standards requirements is closely tied to the specifics of the requirements used in the program and the importance of the standards in determining which households receive assistance, the advantage of a minimum standards allowance apparently rests directly on the extent to which meeting each item of the requirements is itself regarded as critically important. This poses a more severe test for housing requirements than has yet been applied in designing housing programs.

The specificity of responses to minimum standards requirements also suggests that while a minimum rent requirement might not place households in minimum standard housing, it might be more useful in generating more general housing changes usually associated with increased housing expenditures. The minimum rent high housing gap program tested in the Demand Experiment did indeed lead to increases in housing expenditures almost twice as large as those found for unconstrained households (19% as compared to 10%; see Friedman and Weinberg, 1980b: 122-132). That these additional expenditures were not associated with any additional impact on the proportion of households passing various physical standards might be expected. Unfortunately, it appears that they were not associated with any material change in recipient housing at all.

A reasonably stringent minimum rent requirement does force some recipients to spend more for housing. It also, however, does not allow them to take advantage of especially good deals. It is apparent that similar units in similar locations rent for different amounts. Although luck undoubtedly plays a role, to some extent households can find better deals by longer or more extensive search. A minimum rent requirement reduces the incentives to search for the simple reason that if a household finds a good deal it may not meet the minimum rent requirement. As a result, minimum rent high recipients paid more than average for their units. Indeed, the increase in the estimated market value of minimum rent high recipient units was only one percentage point greater than that estimated for unconstrained households (see Friedman and Weinberg, 1980b: 177-178).

Similar problems afflict percent of rent programs. Because these programs do not impose housing requirements, they have the same high participation rate as unconstrained programs. They also lead to larger increases in housing expenditures. A 50% rebate program, for example, would be expected to cause an "average" household to increase its expenditures by 20%, or about 13% percentage points more than an unconstrained program with similar average payments. Again, however, because the program shares in the costs of units, shopping incentives are reduced. On one hand, the program pays for half of any overpayment; on the other, it keeps half of any saving realized by careful shopping. As a result, almost half of the increase in expenditures goes to increased average overpayments. The estimated average market value of units increases only 11%, or 4 percentage points more than a similar unconstrained program (Friedman and Weinberg, 1980a: chap. 5).

These findings substantially reduce the apparent advantages of housing allowances over similar unconstrained programs. The additional housing change generated by allowances is either tightly focused on specific requirements or largely absorbed by changes in recipient shopping behavior. The findings do not, however, undermine the relative cost-efficiency of allowances over other housing programs discussed earlier. In particular, there is no evidence that the effects of other housing programs on recipient housing are any greater than the effects of housing allowances.

The Section 8 existing housing program is so similar to housing allowances in structure that it is difficult to imagine any greater program impacts.[4] In contrast, construction programs and the old Section 23 leased housing programs directly provide program-required housing to recipients. This difference could have important ramifications in terms of program impact on the proportion of households in standard housing. The strong connection between households' normal probability of occupying minimum standards housing and their probability of participating in a minimum standards program is in part definitional. Among households that apply for the minimum standards program, those that already occupy or are about to move to standard housing participate automatically. If the low participation rate among other households reflects difficulty in finding standard units rather than strong cost-preference tradeoffs, construction programs and Section 23 could attract more households that would not normally occupy standard housing. Even if this were the case, however, analysis by Mayo et al. (1980a) indicates that program impacts may still be quite small.[5]

The limited housing impacts associated with housing allowances suggest that allowance programs would have a relatively small impact on housing demand and thus on the supply of low-income rental housing. This may also apply to other housing programs. Again, the similarities between the Section 8 existing housing program and the housing gap form of housing allowance suggest similar effects on housing demand and supply. Even construction programs may not have greater impacts; although they directly increase the supply of acceptable low-income rental housing, unless they draw most of their recipients from households that would occupy substantially substandard housing in the private market, they undermine private demand for standard housing. Thus, they may simply shift households from private to public housing with little eventual impact on the total supply of adequate low-income rental units.

In short, little is known about the effects of other housing programs. What is known is that they are relatively expensive. More could be done, even with existing data, and especially if additional data were collected. In the absence of such analyses, however, it will always be possible for proponents of these programs to justify their additional costs with unsubstantiated claims of added program benefits.

The lack of direct information on the impacts of nonallowance programs indicates several important areas for further research. First, better understanding of housing allowances could allow better predictions of the effects of the differences in program payment calculations and search requirements discussed above. Second, better understanding of the role of search and good or bad deals in determining participation could suggest improved allowance program designs to increase participation among those in substandard housing.

There are also important and unexplained site differences in the estimated effects of both housing gap and unconstrained programs. While many analyses yielded similar results in both sites, there is evidence that these programs yielded much larger changes in housing in Phoenix than in Pittsburgh. The reasons for this are unclear, but they may be part of a more general problem. The estimates of housing changes developed by Friedman and Weinberg (1980b) do not, in many ways, match the estimates of participation developed by Kennedy and MacMillan (1980). The patterns of response match, but the quantitative values have unexplained discrepancies.

The role of household shopping behavior is also relatively unexplored. Analyses by Kennedy and Merrill (1979) and Friedman and Weinberg (1980a) established that shopping is important and that households respond to changes in incentives to shop. Still lacking is an estimate of the overall variation in housing prices or of the extent to which luck or diligence determine how much households actually pay.

REPRISE

The net result of all this seems to be that housing allowances offer a more efficient form of housing program than new construction. At the same time, it may be desirable to retain some government-owned housing, not because of its proven benefits but because of the failure of allowances to reach most households in substandard housing. In addition, it appears that while allowances may be a reasonable vehicle for providing housing assistance to some low-income families, they will have so little impact on the overall demand for housing that they are extremely unlikely to have any substantial effect on urban decay or abandonment. Again, however, the case for the effectiveness of new construction in addressing these problems is purely theoretical and even theoretically tenuous. Finally, given their limited impacts, it is not at all clear that allowances are preferable to expanded welfare support.

Taken together, the Experimental Housing Allowance Program was certainly the largest sustained effort ever made to understand what a housing program does. The experiments have yielded much detailed information about allowances and considerable insight into other housing programs as well. At the same time, they emphasize the limited and often misleading information available on other housing programs and our still rudimentary understanding of housing markets. While there are clear conclusions about relative program costs and effects, many questions remain unanswered.

These unanswered questions are, I believe, in the nature of experimentation and arise in part from initial design decisions. If we had known where our analysis would lead, we would probably have increased the sample size in some treatments, extended the experimental period for some subsets of households, and devoted more resources to a thorough investigation of search behavior.

The most serious small sample size problem arose from a quite different source: We deliberately kept the unconstrained (expanded income maintenance) treatment small on the grounds that the U.S. Department of Health, Education and Welfare was then running several large income maintenance experiments which would, we felt, provide substantial information on the effects of unconstrained income transfers. In retrospect, this was remarkably naive. It is in fact difficult to use data from another study; the data are rarely available when you want them and require learning and adjusting for different situations and data conventions. More important, an analysis focused on housing builds up a large set of ancillary analyses and outcome variables that are unlikely to be produced in experiments focused on work effects. Replicating these in another experiment is often impossible and in any case enormously time-consuming. As a result, our analysis of the effects of expanded welfare payments outside of simple rental expenditures change was more dependent on cross-sectional data than we would have liked. Most fundamentally, however, the real consistency in the housing gap results of the Demand Experiment across the two sites lay in the comparison between housing gap and unconstrained households. For this finding we were completely reliant on the unconstrained sample. The lesson here is straightforward; if an option is really important, it should be included in the design without relying on other data sources.

An increase in housing gap sample sizes would also have been useful, since low participation rates substantially reduced the effective sample and led to a fair amount of noise in individual cells, especially given the relatively small responses involved. This was, however, much less important than the small unconstrained sample. The housing gap sample sizes were generally adequate to establish the central points in the analysis, and no research has ever had enough sample in the end.

The use of a three-year experimental period was an issue during the original design. The designers of the Seattle-Denver Income Maintenance Experiments included a 20-year guarantee in their design. This was not needed here; few people remain eligible, and few, if any, government programs have a stable guarantee for that many years, though an extra year and possibly a five-year group would have been useful.[6] Furthermore, the risks to extended experiments are relatively small—if it becomes apparent that extra years of observation are not in fact needed, the data collection and analysis and even program operation costs associated with the extra years can be cut off by buying out enrollees' remaining payments.

The most important weakness in the design was probably its relative inattention to issues of search. In particular, we could have offered a range of support services ranging from none to individual aid in search to offering a set of specific units that met program requirements. We could also have included a set of standards that could much more often be met by easy repairs and designed our data collection to estimate the cost and labor needed to meet requirements by repairs. These changes in design are, however, the fruits of analysis. We did consider variations in supportive services and physical standards in the design, but at the time the issues involved were not clear enough to dictate well-defined variations or justify the additional sample required.

The unanswered questions also arise in part from the process of experimental analysis. When we started the analysis of the Demand Experiment, we correctly perceived that it ultimately had to involve considerable modeling. Experiments are immediately undertaken to test specific program options, but they must also be viewed as observations on a continuum of options. The interest in a particular program possibility will ebb and flow over the years; given their expense, social experiments must cast a wide net and be viewed not simply as providing a simple yes or no answer to a specific program option but as exploring a range of options and creating a deeper understanding of the basic problems to be addressed.

The final analyses of the Demand Experiment are, in fact, notably model-free. There is enough modeling to provide a basic interpretive framework, to identify covariates, and worry about estimation issues, but the models are loosely specified. We found that our early models were too crude and clearly missed major determinants of behavior. We fell back on the strength of the experimental design—its ability to provide relatively model-free estimates of impact. This is, I believe, quite proper. The first step in experimental analysis is to get the point estimates straight, a process that is hardly mechanical. Statistical issues abound in social experiments despite their experimental design, if only because there are usually quite complicated issues of self-selection in participation and attrition.

These simple point estimates are not enough, however. Policy makers need the stories that go along with them. Do low participation rates reflect financially inadequate offers (from the policy maker's perspective), elimination of the temporarily poor or the more permanently poor, elimination of those with exceptionally good housing deals (who need assistance least) and those who do not care about housing, or of those with the greatest need for assistance? Answering these questions says whether allowances failed to provide support or were targeted to support these most in need. They may also suggest how allowances might be changed to raise or lower a better target participation. Similarly, does the observation that allowances failed to alter existing patterns of racial and economic segregation indicate a need for passive antidiscrimination support, active counseling, or even more active support in search and unit acquisition? This again depends on the mechanisms that maintain segregated housing.

The point is that policy depends not only on simple outcomes but on the much richer stories that explain and describe these outcomes in terms of the motivations and perceptions involved. The analysis of the Demand Experiment is rich in such dynamic stories, but there is no question that much more could now be done to enlarge the stories, develop their specification, and test their adequacy using the available experimental data.

This is a central tension in the design of experiments. Hausman and Wise (1981) are, I believe, absolutely correct in asserting that experimental designs must include enough observations per cell to generate powerful conclusions about specific points—in effect, that the design sample allocation should not rely on a prior model. Indeed, this feeling played a major role in our decision not to follow a Watts-Conlisk (Conlisk and Watts, 1969) procedure in allocating samples within the Demand Experiment; we essentially imposed a fully crossed analysis (no suppression of any identifiable interaction), which leads to equal cell sizes. On the other hand, it should also be clear that the eventual analysis of an experiment will require considerable modeling and estimation of unobserved points. A rich array of treatments is essential to distinguishing alternative models, hence the tension in design between many points with small samples per treatment and fewer points with larger samples per treatment. My last guess, in retrospect, is that the Hausman-Wise suggestion has priority—cells should generally have a large enough sample to provide a good point estimate for the cell (and first-order contrasts). If this leaves too few points for modeling, you may not want to experiment.

Was it all worth it? Evaluating research is notoriously difficult, but the answer seems to be that the experiments easily produced enough findings to pay their cost many times over. The findings of the Demand Experiment on the relative costs per unit benefit of allowances and new construction programs alone would repay the cost of the experiment a hundred times over if applied by the government to existing programs. The more negative findings on participation and housing change at least can save policymakers from an unwarranted reliance on allowances as a universal remedy or as a tool for addressing urban decay or abandonment. It seems clear that most policymakers and researchers, ourselves included, grossly overestimated allowance program participation rates and hence its impacts.

There is still, however, the question of alternatives. Other researchers, faced with the (for social research) enormous expenditures involved in social experiments, tend immediately to think of what they could do with even part of such funds. But before one is carried away by visions of the best, it is well to remember what actually occurred. The public housing program, for example, has been in existence for almost 50 years. While prior studies had certainly noticed its relatively higher costs, these data were not nearly as convincing without the information on housing quality obtained in the demand experiment. Indeed, even though the study of other housing programs was tangential to its main design, the Demand Experiment represents the most extensive investigation of these programs to date.

NOTES

1. Physical adequacy, crowding, and rent burdens in the different programs are discussed in Kennedy (1980: chap. 3) and Mayo et al. (1980a: chap. 4).

2. Figures given in the text adjust for an assumed acceptance rate of 80%.

3. More recently, a matched pair study of realtor discrimination in Boston, conducted by Feins et al. (1981), amplified Vidal's conclusions. Feins et al. found little perception of discriminatory treatment by prospective renters. In fact, realtors showed minority households far fewer units (in nonminority areas) than they offered to nonminority households, limiting minority search patterns and hence helping to maintain racial segregation.

4. There are differences between the two programs. In particular, tenant payments under Section 8 may not exceed 25% of income, and there is a ceiling on total unit rent.

5. Mayo et al. (1980a: 89ff) estimate program impacts on the value of housing services based on cross-sectional comparisons with control households, which are appropriate only if there is no participant self-selection in terms of normal housing consumption. If there is similar self-selection in public housing and minimum standards, the estimates—controlling for participant demographic characteristics and adjusted for certain accounting anomalies in Pittsburgh (in particular, the fact that most private renters pay separately for stoves and refrigerators), would indicate a marginally smaller impact for public housing. If there is no such self-selection in public housing, then comparison with Friedman and Weinberg's (1980b: 169) estimates of actual minimum standards impact would indicate an increase in housing services of 8% more than housing gap minimum standards in each site (see Mayo et al., 1980a: 92; Kennedy, 1980: 207-208). Federal and local costs of newly constructed public housing, on the other hand, were estimated to be almost twice those of allowances (Mayo et al., 1980b: S-5, 114). These are admittedly tenuous calculations.

6. The limited evidence available, while weak, suggests that a longer period would not have changed the results (see Friedman and Weinberg, 1980a: 122-134).

4

The Administrative Agency Experiment

WILLIAM L. HAMILTON

□ THE ADMINISTRATIVE AGENCY EXPERIMENT was created almost as an afterthought to the Demand and Supply experiments, motivated by a complicated political environment. In 1970-1972, as plans for the Experimental Housing Allowance Program were being drawn up, the Nixon administration seemed about to embrace a new housing policy. The general expectation was that the new policy would emphasize the housing allowance concept. In January 1973, the President placed a moratorium on the major existing housing assistance programs, reinforcing those expectations. And in September of the same year, he announced that housing allowances—labeled "direct cash assistance"—would be a major element of the new national housing policy.

The Demand and Supply experiments could obviously provide important information for formulating a national housing allowance policy. But they had some important limitations in the policy environment of the early 1970s. One limitation was their timing: They would not produce results for some years, and a program might have to be implemented as early as 1974.

The fact that the experiments would be operated entirely by research contractors also had drawbacks. Research contractors would not develop a cadre of experienced program managers in public agencies, and such a cadre might be necessary both to generate political support for the housing allowance policy and to provide leadership and training in implementing the program. Another drawback stemmed from widespread allegations that HUD was mismanaging its existing programs, especially public housing. Assurance was needed that the new program could be managed successfully, but "laboratory experiments"—controlled research environments with research organizations in the operational role—might not provide enough assurance.

The policy environment prompted HUD Secretary Romney, when reviewing the research agenda for the Demand and Supply experiments, to insist on a further field test of the housing allowance concept. The result was a plan for the Administrative Agency Experiment. The experiment, or demonstration (terminology varied in the early period), was scheduled to start quickly and to produce findings before the Demand or Supply experiments. Substantively, the experiment would focus on issues of program administration. Public agencies would run housing allowance programs. The agencies would be geographically dispersed; there were eight, spread among seven of HUD's ten administrative regions. Each of these major features of the AAE responded as much to political and operational goals as to a research agenda.

This book is about the research results of the housing allowance experiments, not about their other consequences. Still, because the AAE was formed with such a strong dose of political and operational purposes, a glance in those directions is worthwhile before we turn to the research.

The Nixon administration's drive for direct cash assistance was quickly overshadowed by Watergate. The policy died with Nixon's resignation, less than a year after it had been announced. This eliminated the need for the political constituency that the AAE would have helped build. But the operational mission–developing pragmatic experience to aid in implementing a national program—was ultimately carried out, though not quite as the administration had envisioned. Congress, in the Housing and Community Development Act of 1974, had incorporated what amounted to a modified housing allowance program in the Section 8 Lower Income Rental Assistance Program. As HUD staff established the structures and regulations of the new program, they drew heavily on the experience of the Administrative Agency Experiment. Indeed, the operating manual for the existing housing portion of the Section 8 program was written by people who had been immersed in the AAE, who made direct use of the material and procedures that had been developed in the experiment.

DESIGN

With respect to its research objectives, the Administrative Agency Experiment was directed at three central questions:

(1) Was the housing allowance concept administratively feasible? Might it work in only some kinds of locations, such as those with high housing vacancy rates and small minority populations? Could it be administered by existing housing agencies, or would different kinds of experience be needed?

(2) Were there important administrative options for implementing a housing allowance program? Should agencies be tightly regulated in their administrative process, or should they have the discretion to write their own operating rules?

(3) Could a housing allowance program be administered at a reasonable cost (especially given the reputation of other housing programs for costing much more than their private sector equivalents)? In particular, could a reasonable housing standard be enforced for anything less than prohibitive costs?

The AAE's findings in each of these areas will be summarized below, after a brief description of the project's overall design.

The AAE can be viewed as made up of eight parallel demonstration programs. Eight public agencies were selected to operate small-scale housing allowance programs. Each agency had a few months to enroll between 400 and 900 participants (a target was established for each agency), and each enrolled family could receive housing allowances for up to 24 months.

Four distinct types of agency were chosen. There were two local housing authorities (Salem, Oregon; Tulsa, Oklahoma); two state agencies responsible for housing programs (Springfield, Massachusetts; Peoria, Illinois), two welfare agencies (Bismarck, North Dakota; Durham, North Carolina), and two county or metropolitan governments (San Bernardino, California; Jacksonville, Florida). HUD asked its regional offices to nominate agencies in each of these categories that had demonstrated management competence and could probably succeed in implementing a new program. Selections from the nominees aimed to provide geographic dispersion and reasonable diversity in terms of urban-rural character, size of the poverty population and the minority population, and tightness of the housing market.

All eight agencies operated the same basic program. The housing allowance benefit was computed by the housing gap formula: $P=C^* - .25Y$, where P was the monthly value of the allowance payment, C* was the estimated "typical" local cost for acceptable housing for a family of a given size, and Y was total household income less certain deductions. Only renters could participate. To qualify for payments, an applicant had to occupy housing the agency deemed "decent, safe and sanitary."

Although the agencies all operated the same program, they could exercise great discretion in their administrative procedures. An "agency program manual" specified certain functions the agencies had to perform and offered ideas about possible alternative procedures, but it left the agencies free to choose among the alternatives or invent new ones.

HUD contracted with Abt Associates Inc. (AAI) to monitor the agencies' operations, collect data, and analyze the results of the experiment. AAI collected three major kinds of data. First, operating forms (such as application and enrollment forms) were centrally designed, maintained by all eight agencies, and entered into a central data base. The data base ultimately contained some 73,000 forms concerning over 15,000 applicant households and nearly 6,000 recipient households. Second, AAI carried out surveys and housing evaluations at three points in time for a panel of about 1,200 households. Third, field observers resided on-site at the eight agencies for the first year to record agency operations in detail.

This is not the place for a critical review of the design of the AAE; it has been both criticized and defended elsewhere (Struyk and Bendick, 1981). But it is important to single out one feature of the design that has often been criticized and that is particularly important in understanding some of the research findings. Administrative procedures were not systematically varied. The agencies chose and designed their own procedures rather than implementing procedures specified in a research protocol. This "naturalistic design" will be seen both to provide some strengths and to require some caveats in the findings discussed below.

ADMINISTRATIVE FEASIBILITY

Long before the first research report was written, the Administrative Agency Experiment had resolved the question of whether the housing allowance was feasible. It worked: The eight agencies all succeeded in putting together a program design and an operating structure, in identifying and enrolling beneficiaries, and in making allowance payments.

Hindsight and the Section 8 program make it easy to shrug off this result as obvious. But it was not obvious in 1972. People raised several serious doubts about feasibility, and a close examination shows they were not all wrong. If anything, the doubters simply overestimated the impact of the factors that concerned them.

For example, community acceptance had been a major concern. Proponents and opponents of housing allowances both expected that the recipient's freedom to choose a dwelling unit would mean increased locational mobility. Some opponents had raised the spectre of federally assisted "block busting" and minority "invasion" of the suburbs, and it was feared that this vision would cause local communities to block the program. In fact, HUD did encounter resistance as it sought agencies to run the AAE demonstrations, but the opposition was less fierce than some had suspected and was generally overcome by HUD's assurances and by the small scale of the programs. (HUD encountered more serious problems in seeking locations for the Supply Experiment; some candidate sites were abandoned when HUD could not overcome resistance to a full-scale program.)

One of the biggest advance concerns had been that housing agencies would not be able to manage a housing allowance program—recall that four types of agencies were included in the AAE, and that this was the only dimension to be so explicitly controlled. The worry was that a housing allowance would require some administrative functions that were unfamiliar to housing agencies, such as issuing payments and perhaps counseling families as they sought housing. Although the skill requirements were real, the fear proved almost entirely groundless. Housing agencies were able to acquire people with counseling skills. There were no systematic distinctions in performance across agency types.

The other major concern about feasibility dealt with the housing market. Even some proponents of housing allowances feared the program would not work in tight housing markets and that it would not work for minorities in segregated markets. The first results of the AAE seemed to lay those concerns to rest. The agencies did manage, in a fairly short time, to bring the planned numbers of people into the program; the market did not keep them out.

But there was an exception to the general pattern which, on examination, showed that the concerns were not misdirected. While seven of the eight agencies achieved at least 90% of their targeted numbers of recipients, Jacksonville's agency obtained fewer than 40%. HUD gave the agency another chance. In its second enrollment period, the agency signed up the targeted number of recipients and met its performance goals.

The problem in Jacksonville resulted from the interaction of several factors (see Wolfe and Hamilton, 1977; Holshouser, 1977). The housing stock was poor, which meant that a large proportion of would-be participants could not qualify for payments without moving. The housing market was tight, which meant that finding a new unit was hard. The market was segregated and discriminatory, which meant that blacks started out in worse housing than whites and had a harder time finding acceptable housing. Meanwhile, the Jacksonville agency enforced its housing standard more rigorously than most others in the AAE and provided less help to families searching for housing. As a result, only 42% of the Jacksonville enrollees ever qualified for payments, compared to an average of 77% in the other AAE sites.

Ironically, although Jacksonville's success in its second enrollment period effectively silenced the remaining doubts about feasibility, the problem had not gone away. The agency adopted a strategy of vastly overenrolling, so that even after high attrition there would be enough households to fill the available slots (see Table 4.1). The agency also recruited households that were likely to qualify for payments; 57% of the enrollees in the first period had been black households planning to move, but these made up only 20% of the second enrollee pool. The Jacksonville example thus pointed out the extent to which the housing market could affect the program—and also showed how administrative responses could keep the issue of feasibility from becoming visible.

TABLE 4.1 Enrollment Results in Jacksonville (percentages)

	Jacksonville		*Combined Other AAE Agencies*
	First Round	*Second Round*	
Percentage of recipient target achieved	38	111	100
Success percentages for black households planning to move (recipients as a percentage of enrollees)	20	26	64
Overenrollment as a percentage of recipient target	120	220	130

SOURCE: Wolfe and Hamilton (1977: iii)

ADMINISTRATIVE OPTIONS

If there is a central theme in the AAE findings, it is that alternative methods exist for carrying out administrative functions and that the choice among alternatives makes a difference.

For some administrative functions, the analysis identified two or more procedures or methods by which the function could be performed. No two agencies carried out identical activities, even when they were judged to be using the same method. But classifying agencies by method produced workable groups, and analysis showed smaller differences within groups than across them. Moreover, the administrative methods resulted from the natural development of local strategies, not from external direction. This at least opens the possibility of central regulation: It indicates that HUD could require local agencies to carry out particular methods and achieve reasonable cross-agency conformity of procedures.

The AAE analysis considered ten direct functions (involving contact with clients) and six indirect functions (providing direction and resources for conducting the direct functions). All the direct functions had discernible alternative methods. Some had two or more method dimensions that could vary independently; supportive services, for example, could be provided in an individual or a group format, and it might or might not involve a substantial emphasis on ad hoc response to individuals' problems. In some cases, like outreach, methods had to be described in terms of a continuum of intensity without clear dividing lines. Such a definition makes regulation more difficult. Nonetheless, the direct functions generally offered usable handles for policy.

AAE analysis did not reveal clear alternative methods for the indirect functions, however. Agency procedures seemed to represent eight variations on a common theme rather than two or more distinct alternatives. This result may stem partly from limitations of the research; the direct functions received much more intensive analytic effort than the indirect functions. But because indirect functions in social program administration absorb a large share of administrative resources—about half, in the AAE—it will be important for future research to extend the search for policy leverage in this area.

Four of the direct functions were found to have important influences on a housing allowance program's cost and/or its effectiveness. Much of the analysis therefore focused on these four functions—outreach, supportive services, housing inspection, and income certification.

OUTREACH[1]

The AAE agencies had to introduce a new social program to its intended beneficiaries. The analysis of procedures for making the program known shed some light on the low-income population's response to a new program, as well as on the effectiveness of outreach.

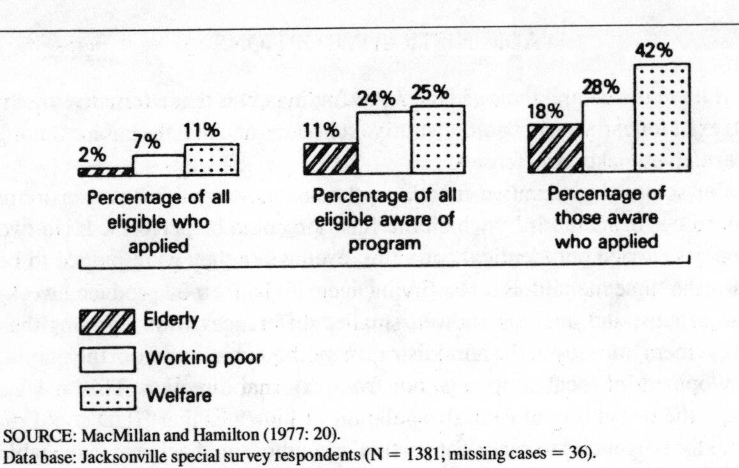

SOURCE: MacMillan and Hamilton (1977: 20).
Data base: Jacksonville special survey respondents (N = 1381; missing cases = 36).

FIGURE 4.1 Application Rates for Elderly, Working Poor, and Welfare Recipients in Jacksonville

The most striking feature of the application patterns in the AAE was the consistent difference in application rates of groups defined as the elderly, working poor, and welfare populations. Compared with their proportions in the eligible community, elderly and working poor households were underrepresented among applicants to all eight agencies, while welfare households were consistently overrepresented.

The consistent patterns for low-income elderly, working poor, and welfare populations suggest that the groups may have distinctive characteristics that influence their response to new social programs and perhaps their participation in established programs. The elderly, for example, were less likely than the nonelderly to be aware of the program, and even those elderly who were aware were less likely to apply (see Figure 4.1). Several factors may influence this pattern. The elderly did not appear to lack formal sources of information (their exposure to media was about the same as that of the nonelderly), but their informal communications networks may be weaker or less effective. The elderly may be more reluctant to apply because older people are less likely to accept new programs, or their reluctance may be due to a stronger value orientation against accepting public assistance.

The differing application rates of the nonelderly working poor and welfare populations likewise merit further investigation. As the Jacksonville survey indicated, welfare households were somewhat better informed about the program and substantially more likely to apply to it than the working poor. The greater awareness of welfare households might result simply from greater interest; the survey indicated no significant difference between the sources of information for welfare and working poor groups. On the other hand, the welfare population's contacts with service agencies or their informal communications networks could bring them word of a new program more quickly than the working poor receive it.

The AAE analysis also showed that the program administrator is not entirely at the mercy of natural response patterns in the population. Outreach works—at least to a degree. The analysis revealed what most practitioners would accept as common sense: that outreach produces applications to a new program. Increases and decreases in the intensity of outreach efforts, as measured by agency expenditures, led quickly to increases and decreases in the number of applications. The finding held for all eight agencies, across a substantial variety of locally designed outreach approaches. An interesting implication is that the general intensity of outreach can be effectively controlled through the budgetary process; more complicated regulation of procedures is not required.

Outreach through referral channels and through the media were the two main outreach methods under the administrators' control. These accounted for 60% of the AAE applicants, with the remainder reporting that they heard of the program through word of mouth. By varying their emphasis on the two major outreach methods, agencies could influence the pattern of applications. Media outreach, although more costly, was substantially more effective in reaching the underrepresented elderly and working poor populations. But even when targeted directly to the elderly or working poor, media outreach rarely produced application rates equal to those of the welfare population.

SUPPORTIVE SERVICES[2]

A troublesome feature of the housing allowance concept, pointed up in the Kaiser Committee discussions and the public debate, was the possibility that it might fail to serve those most in need. People currently living in bad housing would have to succeed in the housing market—locate acceptable housing and sign a lease with the landlord—in order to qualify for payments. But they might not be able to succeed in the market without help. Indeed, their current bad housing might be the result of discrimination, a tight market, inadequate searching skills, or other factors that would keep them from succeeding in the context of a housing allowance program as well.

The debate produced a great variety of actions suggested for an administering agency, ranging from day care to legal help, training to transportation. But it produced no consensus. Hence, the AAE agencies were instructed that they must make available legal assistance for discrimination cases and a minimal level of information. Beyond that, they were free to provide services or not, as they chose.

Their choices varied greatly and on numerous dimensions—so much, in fact, that a major task of the analysis was to identify a small set of dimensions on which to characterize the differences. Initially the most useful classification seemed to be one that specified the content of the services offered (e.g., information about available housing, assistance in negotiations with landlord) or the format of service (whether audiovisual aids were used to communicate information). Another analytic tack focused on the philosophy of agency service policies, identifying laissez-faire, casework, and enabling approaches. Ultimately, the most powerful distinction proved to be whether or not the agency provided a substantial amount of "responsive" service—that is, service given on an individual, ad hoc basis as enrollees asked for help.

The availability of responsive services turned out to improve substantially the chance that an enrollee would qualify for payments—in some circumstances (see Table 4.2). In loose housing markets, services made no difference. In tight housing markets, for households that had to move to qualify for payments, and especially for black households, services were important. Black households planning to move in tight markets stood a 26% chance of qualifying for payments if their agency did not provide substantial responsive services. Their chances improved to 65% with services. Responsive services were thus proven an important mechanism for improving equity in a program that depends on the individual's interaction with the housing market.

HOUSING INSPECTIONS[3]

In the early stages of designing the housing allowance program, HUD planners worried extensively about how to enforce the housing standard. The reason was cost. Estimates from municipal code enforcement agencies had suggested that costs might be as high as $100 per inspection. Enrollees

TABLE 4.2 Proportion of Enrollees Becoming Recipients Under Various Service and Housing Market Conditions, by Race (percentages; sample sizes in parentheses)

		Moving Plans				
		Move		*Stay*		
Market Conditions	*Level of Services*	*Black*	*White*	*Black*	*White*	*Total*
Tighter	High (Springfield, Durham)	65 (290)	63 (336)	77 (223)	77 (470)	70 (1319)
	Low (Peoria, Jacksonville)	26 (219)	54 (389)	53 (92)	76 (552)	51 (1252)
Looser	High (Tulsa)	82 (132)	84 (211)	84 (38)	88 (479)	86 (860)
	Low (Salem, San Bernardino, Bismark)	78 (67)	81 (779)	85 (40)	89 (1063)	85 (1949)
Total		47 (708)	69 (1715)	71 (393)	84 (2564)	71 (5380)

SOURCE: Holshouser (1977: 22).

Data base: Enrollees (N = 7719; missing cases, 376). Most missing cases are other ethnic groups or those who were undecided about their plans to move.

might need two or three units inspected before finding one that met standards, which would be an intolerable multiple of that unit cost. But less costly procedures might result in many subsidies for substandard housing, defeating the program's intent and setting up a public relations nightmare.

The AAE agencies were accordingly encouraged to find less costly means of housing inspection that would still guard against supporting substandard housing. They did, adopting varying approaches to the question of who collected data on the housing unit. Some agencies used specialized housing inspectors like those who enforce the municipal codes. Some used "generalist" agency staff, and some relied on enrollees to fill out a form describing the key characteristics of their chosen units.

Analysis revealed the expected tradeoff between accuracy and cost. Professional inspectors were most expensive but produced enough information to make accurate judgments about substandard units in 87% of the cases. Participant inspections involved the lowest cost and the highest risk, providing enough information to reject only 36% of the substandard units in the sample.

The finding on costs was in the expected direction, but surprising in absolute amount. The average cost of the professional inspection was only about $36 per enrollee, far below most early estimates. The low figure resulted in part from the fact that AAE inspections were a bit simpler and entailed much less complicated follow-up procedures than is typical of code enforcement. Equally important, the median number of inspections per enrollee was 1.07. Many did present two or three or more units for inspection, as anticipated, but these were offset by a large number of enrollees who dropped out of the program without requesting even a single inspection. Thus, even though the the participant inspection was considerably cheaper, at about $5 per enrollee, the professional inspection turned out to be an affordable means of reducing the risk of substandard units in the program.

INCOME CERTIFICATION[4]

The housing allowance payment in the AAE was based on participants' household size and their income. Because faulty data on these points could cause mispayments, the agencies had to certify the accuracy of the information. This certification required some review that could spot errors. It could involve verifying the information by requesting documents or checking with knowledgeable third parties, or it could simply involve an additional review and sworn statement by the applicant.

The certifications revealed a high potential for error at the time of application. Few errors were found in the data on household size. But the agencies had to adjust the initially reported income data in more than half the cases. Had these adjustments not been made, the average participant's annual benefits would have included an error (either overpayment or underpayment) of $116, or about 13% of the average benefit.

Payment error tends to be equated in the public mind with dishonesty—attempts by participants to obtain benefits to which they are not entitled. The consequent assumption is that eliminating error will substantially reduce public expenditures. In the AAE, however, agency efforts to adjust income data would have reduced the average payment by less than $1 per year (for those who actually did receive payments). The reason for this small effect was that the agencies made almost as many adjustments that increased payments as adjustments that reduced them. Although the source of error cannot be measured in the AAE data, the pattern appears to be dominated by honest error. Agencies reported a handful of cases of deliberate misreporting, and additional cases doubtless went undiscovered. But the major contribution of agency certification was not to save money by identifying instances of fraudulent behavior but to allocate program resources more equitably among eligible households.

VERIFICATION

Checking participant-reported income data by examining documents or contacting knowledgeable third parties proved an important means for avoiding error in the AAE. Any procedure for reviewing income data resulted in a substantial number of adjustments to participants' reports, but verifying the information through documentary or third party evidence produced markedly more adjustments than merely reviewing the information with participants. On average, spending an additional $4-$6 for verification avoided $28-$34 in annual payment errors, including both overpayment and underpayment errors.

The high rate of error avoidance and the low marginal cost seem to argue that income data at application should normally be verified. But there is an important qualifier to the finding. Although verification did predictably reduce errors, it did not yield a predictably greater reduction in total payment expenditures. In other words, universal verification would actually increase total government expenditures, because the administrative cost would not be offset by payment reductions.

ADMINISTRATIVE COST

Estimating the cost of administering a national program stood prominently among the AAE objectives when the project began.[5] But policy bypassed the research, and by the time total cost estimates became possible they had also become less relevant. They are useful now mainly to establish a perspective. With a central estimate of about $240 per recipient per year (1974 dollars), the housing allowance program would have administrative costs similar to those for welfare (AFDC). Not surprisingly, Section 8 administrative cost estimates are in the same range.

In the absence of an ongoing program, the more interesting findings concern particular components of administrative cost and the ways they can be influenced by policy. For example, the policymaker's choice among alternative procedures can make a substantial difference to administrative costs. Suppose an agency chose to use the most expensive procedures in each case where the AAE analysis identified alternatives. Its estimated cost would be about $285 per recipient year (assuming a constant indirect cost rate). An agency choosing all of the cheapest procedures would have a cost of about $147. In other words, if all of the agencies in a national program were using the most expensive procedures, a HUD order to use the least costly methods could approximately halve administrative cost.

Another policy handle connects to a cost component dubbed "attrition cost." To bring families into the program, AAE agencies had to perform outreach, accept applications, screen out those obviously ineligible, place the remaining applicants in a pool (assuming more applications than currently available slots), select from the pool, certify the income of the selected applicants, enroll those found eligible, and determine the acceptability of housing units presented by enrollees. Each step costs money. A family that went through the whole process to qualify for payments incurred an estimated $133 of administrative costs. If the family then failed to qualify for payments, the $133 was an attrition cost.

Program budgets and efficiency assessments usually consider a total cost per recipient. Because of attrition cost, the more applicants who fail to become recipients, the higher the recorded administrative cost per recipient. The housing allowance program in the AAE had two features that could produce substantial attrition. The number of slots was limited, so any excess of applicants over slots could cause attrition. Enrolled households had to find acceptable housing, and we have already discussed the substantial attrition that could occur at this point. As a result, attrition costs accounted for about a third of all expenditures for program intake.

A policy to control attrition may thus reduce administrative costs. This raises the intriguing possibility that the policymaker could choose an expensive procedure to hold down attrition but thereby reduce average costs per recipient. This possibility is illustrated in Figure 4.2. Responsive services to enrollees, as we have seen, increased their chances for qualifying for payments—but only for relatively high-risk groups. Providing responsive services was a more expensive alternative in loose housing markets or with low-risk populations. But in a tight housing market with an enrollee population containing many black households and households that had to move to qualify for payments, the analysis indicates that the costly responsive services would yield a 15% *lower* average intake cost.

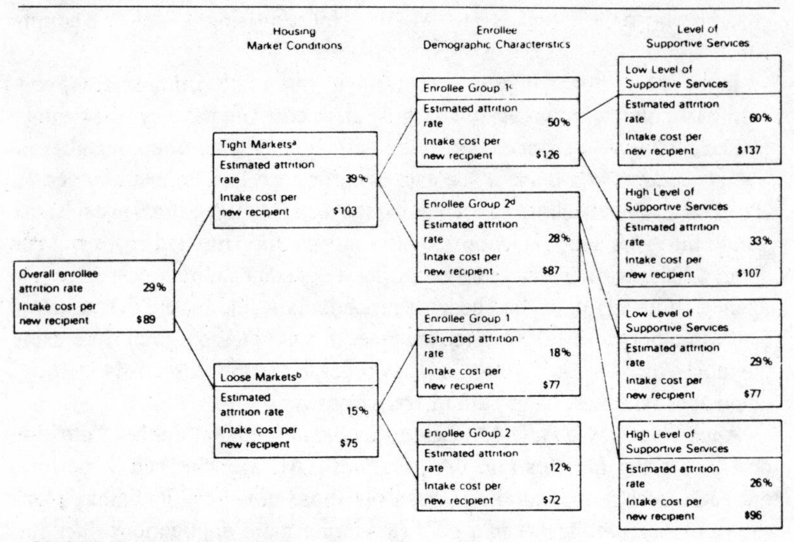

SOURCE: Hamilton et al. (1977: 75).
NOTE: Enrollee attrition rate = (enrollees − recipients) ÷ enrollees.
a. Tight market sites are Springfield, Peoria, Jacksonville, and Durham.
b. Loose market sites are Salem, San Bernardino, Bismarck, and Tulsa.
c. Modeled after the Jacksonville enrollee population; includes 66% black enrollees, of whom 92% plan to move and 8% plan to stay; and 34% white enrollees, of whom 33% plan to move and 67% plan to stay.
d. Modeled after the Bismarck enrollee population; includes 100% white enrollees, of whom 29% plan to move and 71% plan to stay.

FIGURE 4.2 Estimated Enrollee Attrition Rates and Direct Intake Costs

A LATE PERSPECTIVE

One cannot pretend to have summarized in so few pages the work of many people over several years[6]; the examples cited above are intended to give a sense of the dominant directions of the AAE analysis and the general nature of the findings. Judging the contribution of this first social experiment in program administration is even more difficult. The policy changes of the 1970s meant that the AAE research results, like the rest of EHAP, would have their effects indirectly, through becoming part of the literature and understanding of social program operations, rather than in the immediate use of findings to formulate policy. Perhaps the most interesting test will be to see whether the 1980s—which appear destined to emphasize cost containment and administrative efficiency, but not social research—will bring about projects that follow some of the paths traveled in the Administrative Agency Experiment.

NOTES

1. See Macmillan and Hamilton (1977).
2. See Holshouser (1977).

3. See Budding (1977).

4. See Dickson (1977).

5. For AAE analyses of administrative costs, see Hamilton et al. (1977a, 1977b) and Maloy et al. (1977).

6. Summaries of AAE research are included in Hamilton (1979) and Hamilton et al. (1977b).

Part II

Eligibility and Participation

☐ PART II deals with eligibility and participation issues. Chapter 5, by Grace Carter, Sinclair Coleman, and James Wendt, presents the findings on participation in the full-scale housing allowance programs run as part of the Supply Experiment. Participation in those programs was open to all eligible households in the two HASE sites (Brown County, Wisconsin and St. Joseph County, Indiana) who chose to apply and comply with the housing requirements. Yet, after three full years of program operation, the steady state rate of participation appeared to be that only a third of all eligible households were receiving allowances. The chapter explores two issues related to such low participation rates: (1) Why did some households and not others participate? (2)Who participated (received a housing allowance)?

The authors attribute the low participation rate to three factors—lack of knowledge about the program, low level of perceived need (associated with expecting a low level of benefits or a short duration of eligibility), and the perceived high cost of participation (when moving from or repairing substandard housing was required). Then they examine the causes of nonparticipation by examining the acquisition of program knowledge, the decision to enroll, and the response of households whose housing failed the program's minimum standards. Households in better housing were more likely than those in poor housing to enroll in the program and receive payments. Elderly singles and single parents had the highest participation rates. Couples with children participated less than the other groups because of their larger household size and shorter duration of eligibility.

Chapter 6, by Stephen Kennedy and Jean MacMillan, summarizes their findings on participation in the Demand Experiment. HADE tested 17 different housing allowance variants. Thus, because households were randomly selected to receive allowance offers (or to participate as controls), it was possible to determine how varying program parameters affected participation.

The central finding of their analysis was that the imposition of housing requirements sharply reduced participation among households that would not normally live in housing meeting those requirements. Because of this, demographic groups that are less likely to meet requirements initially (minority households, large households, and households with very low incomes) are less likely to participate in a housing allowance program than in a similar unconstrained program. These patterns can, however, be changed by changes in the allowance payment—increasing the average payment level (guarantee) or changing the benefit reduction rate.

Chapter 7, by William Holshouser, discusses the role of supportive services in encouraging participation, based on results of the AAE. Supportive services—program information, counseling, legal services, transportation, child care, assistance in finding units or negotiating with landlords—are presumed to help program participants make better use of their allowances. The need to experiment with supportive services arose because many service advocates were convinced that low-income tenants armed only with housing allowances could not succeed in the housing market. They were convinced that supportive services were needed, at least for an interim period, presumably until the poor gained more experience. On the other hand, there was no evidence that previous attempts to provide counseling and other services was ever helpful.

Experimentation with supportive services was possible within the eight sites of the AAE. Formal services (standardized services presented to enrollees shortly after their enrollment) were found to have little effect on success rates. Responsive services (provided to individual enrollees with problems as need arose), however, had a tangible positive effect on program outcomes. Further, responsive services provided the most help to those with the largest problems—black households and those attempting to find units in tight housing markets.

5

Participation Under Open Enrollment

GRACE M. CARTER, SINCLAIR B. COLEMAN, and JAMES C. WENDT

☐ THE HOUSING ALLOWANCE PROGRAM operated by the Supply Experiment (HASE) was open to all eligible households in Brown County (Green Bay) and St. Joseph County (South Bend) that choose to join and comply with the housing standards. After three years of operation the program appeared to have reached a steady state of participation; yet only a third of all eligible households were receiving allowances. This chapter explores two issues related to such low participation rates: (1) Why did some households and not others participate? (2) Who did participate—that is, which households received allowances?

The reasons some households did not participate fall into three basic groups: First, some households lacked knowledge of the program. Second, some households had a low level of perceived need, usually because they expected low benefits or short durations of eligibility. Third, for some households the costs of participation (primarily the cost of obtaining adequate housing) were too high. The implications of these participation rates for a national program depend on how much each of these explanations contributed to the low participation in the experimental sites. In the next section we analyze the cause of non-participation by examining the acquisition of program knowledge, the decision to enroll, and the response of households to a failed housing evaluation. To answer the second question, we analyze patterns of participation among the eligible households in a subsequent section.

The remainder of this introductory section contains a brief description of the data sources, the eligible populations of each site, and program growth.[1] Two parts of the HASE data base are used for this analysis: the four annual household surveys and five years of housing allowance office (HAO) administrative records. Since the surveys were administered to the same dwellings

---

Reprinted by *permission of The Rand Corporation.*

TABLE 5.1 Number of Households and Fraction Eligible, by Tenure, Life Cycle Stage, and Race of Head: 1977

	Brown County			*St. Joseph County*		
	Number of Households			*Number of Households*		
Household Type	*Total*[a]	*Eligible*	*Percent Eligible*	*Total*[a]	*Eligible*	*Percent Eligible*
			Renters			
Life-cycle stage:						
Elderly couple	489	266	54.4	747	277	37.0
Elderly single	1,226	771	62.9	1,797	1,271	70.7
Single parent	1,693	1,296	76.6	2,370	1,925	81.2
Young couple, young children	2,732	845	30.9	2,298	642	27.9
Other couple	2,855	511	17.9	2,935	529	18.0
Nonelderly single	5,347	0[b]	0.0[b]	5,472	0[b]	0.0[b]
All stages	14,342	3,689	25.7	15,619	4,644	29.7
Race of head:						
White, non-Hispanic	c	c	c	12,889	3,211	24.9
Other	c	c	c	2,730	1,433	52.5
All races	14,342	3,689	25.7	15,619	4,644	29.7
			Owners			
Life-cycle stage:						
Elderly couple	3,390	1,087	32.1	7,989	2,399	30.0
Elderly single	2,615	1,007	38.5	6,792	4,193	61.7
Single parent	1,778	677	38.1	3,772	1,876	49.7
Young couple, young children	9,612	858	8.9	12,981	1,236	9.5
Other couple	13,603	594	4.4	17,699	1,052	5.9
Nonelderly single	2,732	0[b]	0.0[b]	7,036	0[b]	0.0[b]
All stages	33,730	4,223	12.5	56,267	10,755	19.1
Race of head:						
White, non-Hispanic	c	c	c	51,446	9,297	18.0
Other	c	c	c	4,821	1,458	30.3
All races	33,730	4,223	12.5	56,267	10,755	19.1
			Total			
Life-cycle stage:						
Elderly couple	3,879	1,353	34.9	8,736	2,676	30.6
Elderly single	3,841	1,778	46.3	8,589	5,464	63.6
Single parent	3,471	1,973	56.9	6,141	3,801	61.9
Young couple, young children	12,344	1,703	13.8	15,279	1,878	12.3
Other couple	16,458	1,105	6.7	20,633	1,580	7.7

TABLE 5.1 (Continued)

	Brown County			*St. Joseph County*		
	Number of Households			*Number of Households*		
Household Type	*Total*[a]	*Eligible*	*Percent Eligible*	*Total*[a]	*Eligible*	*Percent Eligible*
Nonelderly single	8,079	0[b]	0.00[b]	12,508	0[b]	0.0[b]
All stages	48,072	7,912	16.5	71,886	15,399	21.4
Race of head:						
White, non-Hispanic	c	c	c	64,336	12,508	19.4
Other	c	c	c	7,550	2,891	38.3
All races	48,072	7,912	16.5	71,886	15,399	21.4

SOURCE: Carter and Balch (1981: Table 4.1)

a. Resident landlords and residents of subsidized housing are excluded.

b. Single persons under 62 are classified as ineligible, although a few were eligible due to handicaps, disabilities, or residential displacement.

c. Not estimated for Brown County, where nearly all residents are non-Hispanic whites.

each year, many households were interviewed repeatedly. Some surveyed households were encouraged to apply to the program by questions about it, thus biasing the survey data on participation. But because the same data include many households that were not interviewed previously (mainly households that had moved within the preceding year), we can control for such bias in our analysis.

We estimated the sizes and composition of the household populations eligible for housing allowances in each county for each year covered by the surveys. This was done by applying the income, asset, and family composition rules to each complete survey record and inferring the population statistics as described in Carter and Balch (1981).

In Table 5.1 we present estimates of the populations eligible for housing allowances in 1977, classified by housing tenure, life cycle stage, and (in St. Joseph County) race of head. As one might expect, eligibility rates among renters are much higher than among homeowners; the latter are typically more prosperous. However, homeowners are also more numerous in the population, and their low eligibility rates nonetheless yield a majority of the eligibles in both sites.

Renter elderly singles and single parents have much higher eligibility rates than any other groups. Among the elderly, the lower eligibility rates for couples than for singles are due to both a greater amount of non-social security income and larger social security payments relative to standard rent. In St. Joseph County, households headed by nonwhites are twice as likely to be eligible as those headed by whites.

During the three years covered by our four surveys, the number of eligible households changed very little in either site: in Brown County it increased by 7%, matching the growth rate of the household population; in St. Joseph County, where the population was not growing, the number of eligibles did not change significantly.

Despite the absence of large change in the number of eligibles, there were many changes in the eligibility status of individual households due to changes in household income, marital status, household formation and dissolution, and a small amount of migration into and out of each site. About 20% of the eligible households became ineligible each year. Another 10% dissolved or moved away (our data do not distinguish between these two effects), leaving just 70% of each year's eligible pool still eligible a year later. (See Carter and Balch, 1981, for details.)

Participation in the allowance program grew rapidly during the first two years of program operation but leveled off by the end of the third program year and remained roughly constant for the last two years of the experiment.[2] The heavy advertising campaigns in each community appeared to have accomplished their purpose. By the end of the third year 80%-85% of the eligible population (depending on the site) knew enough about the program to describe some of its details, and we expect that further spread of program knowledge was limited (see below). The household survey is the source of our information about eligible households that did not enroll; the last wave of surveys approximately coincides with the end of the third year of program operation.[3] However, because of the leveling off of enrollment and knowledge, we believe the effects of program startup had disappeared by the time of the last survey.

Although the number of households in the program stabilized after the third year, their identity continued to change; each year about one-third of those receiving payments left the program and were replaced by others. This turnover among recipients mainly reflected turnover among the eligible population.[4] By the time the program reached steady state at the end of the third year, only about one-third of the eligible households were receiving payments. To understand the cause of this low participation rate, we next analyze the steps in the participation process.

DETERMINANTS OF PARTICIPATION

Before a household could receive payments, it first had to learn about the program and then enroll in it. After enrollment, the household's housing unit was inspected. If it met program standards, payments began immediately. Otherwise the household had to either repair the dwelling, move to an adequate one, or forego the allowance payment. Because we expected that the determinants of behavior would differ at each step, we decomposed our own analysis into three steps: (1) acquiring knowledge, (2) enrollment, and (3) response to dwelling failure. Each of these is discussed separately below.

TABLE 5.2 Progress Through the Steps of the Participation Process

		Percentage of Households in Group A that are in Group B			
		Brown County		*St. Joseph County*	
Group A	*Group B*	*Renters*	*Owners*	*Renters*	*Owners*
Eligibles	Have knowledge	85	78	85	85
Have knowledge	Enroll	77	50	75	43
Enroll	Receive payments	83	87	72	86
Receive payments	Currently receiving payments	82	84	83	87
Eligibles	Currently receiving payments	45	28	38	27

SOURCE: Carter and Wendt (1982: Table 3.1)

Table 5.2 shows the fraction of households that successfully completed each step in the participation process. As we have seen, the eligible population always contains many newly eligible households. Because it takes some time for these households to decide to enroll, and for enrollees to receive payments, some households will eventually receive payment but are not yet doing so. The proportion of those who are currently eligible and are ever recipients that are currently receiving payments, shown in row 4 of Table 5.2, gives the combined effect of turnover and delays. The last line of the table shows steady state participation rates.

PROGRAM KNOWLEDGE

Survey respondents were considered knowledgeable on the basis of their answers to two questions: "Have you heard about the housing allowance program which has been introduced in (site)?" If they responded yes, they were then asked, "Suppose somebody asked you about the housing allowance program—how would you describe it?" The knowledge variable was then constructed by assigning a zero to respondents who admitted a lack of knowledge or if they claimed knowledge but could furnish no details of the program, and a 1 if they claimed knowledge and could furnish details.[5]

Table 5.3 shows the percentage of the sample of broad categories of tenure, income, minority status, and age that knew about the program at waves 3 and 4 in each county. There were no large differences within the first three pairs of groups. However, there was a large difference between the elderly and nonelderly in Brown County at both waves, while in St. Joseph County that difference had disappeared by wave 4. In both sites the elderly were the

TABLE 5.3 Percentage of Sample Households Knowledgeable After Years 2 and 3

	Knowledgeable Households (%)			
Household	*Brown County*		*St. Joseph County*	
Characteristic	*Year 2*	*Year 3*	*Year 2*	*Year 3*
Tenure				
Owner	77	82	75	84
Renter	83	86	82	87
Income level				
Under $5,000	75	81	80	84
$5,000 or over	86	87	84	87
Minority status				
White non-Spanish	80	83	78	86
Other	85	89	90	87
Age of head				
Under 62	89	87	88	86
62 and over	61	77	66	85
Total	81	84	81	86

SOURCE: Wendt (1982: Table 6).

TABLE 5.4 Logit Model of Knowledge Acquisition

	Brown County		*St. Joseph County*	
Variable	*Coefficient*	*t-statistic*	*Coefficient*	*t-statistic*
Constant	−1.463	3.33	−0.492	1.18
Previous interview	1.022	4.11	0.535	2.22
Interview X Wave 2	−0.948	2.76	−0.727	2.14
Wave 2	−0.153	0.56	−0.647	2.42
Wave 3	0.007	0.03	−0.008	0.04
Rent expense	0.008	3.39	0.007	3.53
Homeowner expense	−0.003	1.23	0.001	0.45
Household size	0.084	1.06	−0.046	0.66
Parents	0.999	3.59	0.880	3.30
Childless nonelderly couples	0.367	1.29	0.338	1.17
Income ($ 000)	−0.085	2.80	−0.051	1.52
Owner	1.493	2.75	0.966	2.00
Information from				
TV	0.446	2.33	0.257	1.23
Radio	0.237	1.39	−0.013	0.07
Newspapers	0.642	3.73	0.515	2.93
Network	0.102	0.44	0.268	1.29
Government aid	0.425	3.73	0.570	2.81

SOURCE: Carter and Wendt (1982: Table A.1).

NOTE: Sample sizes were 1007 in Brown County and 956 in St. Joseph County. Chi-square statistics were 137.7 and 109.4, respectively.

hardest group to reach and learned most slowly, but a special outreach program in 1977 targeted to the elderly in South Bend appears to have been successful.

To gain a better understanding of how information is acquired, we developed and estimated a model of information acquisition. The data are from eligible households in the waves 2, 3, and 4 surveys in each site.[6] The model was fit using maximum likelihood logit regression. The form of the model is:

$$\ln\left(\frac{P}{1-P}\right) = \alpha + \beta X \qquad [5.1]$$

where P = probability that a household knows of the program
X = vector of explanatory variables; and
α, β = coefficients.

We expected that those with a greater perceived need for the program would pay more attention to available information about it and that perceived need for the program would increase with lower income, larger household size, and greater housing expenses.[7] However, only renters' housing expenses and income appear to affect the acquisition of program knowledge (see Table 5.4).

The previous surveys were related to the respondent's knowledge about the program. We expect that the long interview made respondents more attentive to any other information about the survey in the environment, including the connection between the survey and the allowance program.[8] Thus the effect of previous interviews is greater at waves 2 and 3 when the program was more widely known.

In addition to "selective attention," the acquisition of knowledge depends on exposure to relevant information. To measure exposure, we used four binary variables indicating the household's most important source of information about local affairs[9]—newspapers, television, radio, and friends, neighbors, or relatives. Employees of welfare agencies and unemployment insurance programs were informed about the allowance so they could pass this information on to clients. This strategy appears to work, since those receiving aid from these sources were more likely to know about the program. Dummy variables for survey wave are used to measure the length of time the household was exposed to program information. Although households at wave 2 were less likely to know about the program, there was no noticeable difference between households at wave 3 and wave 4.

The remaining variables in the model distinguish household types: dummy variables for owners, for households with children, and for other nonelderly households. Households with children were more likely to know about the program, possibly because their childrens' activities brought them into greater contact with other members of the community.

THE DECISION TO ENROLL

Roughly three-quarters of renter households and half of owner households who were eligible and aware of the program enrolled (see Table 5.2). However, an average of about 6 months elapsed between when a household became eligible and aware of the program and when it enrolled. We could find no relationship between the costs and benefits of program participation and the length of the enrollment delay. Consequently, we believe that the timing of enrollment is caused by a random process that brings the existence of the program to the forefront of the household head's mind.

Our analysis of the decision to enroll is based on households that were eligible and aware of the program at wave 3 or wave 4.[10] The independent variable is whether the household enrolled at any time during the first five program years. We use a logit model with the independent variables describing the cost and benefits of enrollment.

In exploratory analysis we found that the same model described behavior in both sites, with all differences between sites being captured by a single

TABLE 5.5 Logit Equation for the Probability That an Eligible Household Will Enroll in the Program

Variable	*Coefficient*	*t-statistic*	*Values Compared*	*Change in Percentage Enrolling*
Monthly benefit (1n)	.40	3.24	40-102[a]	7
Liquid assets (1n)	−7.57	2.55	0-944[a]	−10
Duration of Eligibility (1n)	.43	4.52	1-12[a]	16
Moved in last year	1.98	3.96	no to yes	3
Housing expenses per room:				
Renters—not moved	3.32	3.51	26-48[a]	7
Renters—moved	−1.25	1.29	34-50[a]	−3
Owners	−.40	.26	26-48[a]	−2
Persons per bedroom:				
Renters	−.35	.64	.25-1.0	−2
Owners	−1.61	1.57	.25-1.0	−23
Parents	.49	1.84	no to yes[b]	11
Childless nonelderly	−.60	1.88	no to yes[b]	−12
Owners	1.92	2.52		
Against aid	−.64	2.86	no to yes	−13
Brown County	.77	4.77	no to yes	16
Household size	−.04	.41	2-4	−3
Constant term	−3.60	4.01		

SOURCE: Wendt (1982: Tables 12 and A.4).
NOTE: Sample size=981; chi-square=166.4 with 14 degrees of freedom.
a. These values are the 25th to the 75th percentile within the sample.
b. Compared to the elderly default group.

dummy variable. The set of all interaction terms yielded a chi-square statistic of 13.1 with 14 degrees of freedom. Consequently, we report here only the results from pooling the data across sites.

The results are reported in Table 5.5. In order for the reader to grasp more quickly the relative magnitude of effects of each variable, Table 5.5 also shows how the probability that a household will enroll changes when each independent variable covers a typical range and all the other independent variables are held at their sample means. The variables are self-explanatory, except for expected duration of eligibility. This was derived from a logit equation that modeled the probability that an eligible household would become ineligible during a 12-month period as a function of occupation and income sources. We then assumed that duration of eligibility follows a negative exponential distribution within each occupation and income source group, so that the expected duration of eligibility can be calculated for each group.

Enrollment increases with the monthly benefit the household will receive. This probably reflects both the actual incentive value of the dollars to be gained by joining the program and the fact that households with a greater allowance entitlement felt a greater need for the program because of their greater poverty. The second interpretation is strengthened by the fact that the enrollment rate declined with the amount of liquid assets possessed by households; these assets did not affect the benefit amount but should have affected the household's sense of need for the program.

A short duration of eligibility decreases enrollment for two reasons. First, such households will receive payments for only a short time and thus have less to gain by joining the program. Second, even if they would like to join the program, they may become ineligible before they get around to joining it.

A major cost associated with program participation is the cost of moving or repairing a failed dwelling. Thus we expected that those living in the worst dwellings would join the program less frequently. However, enrollment increased with increasing rent only for renter households that had not moved in the year preceding the survey. In addition, the extent of crowding[11] did not significantly affect enrollment. If households were aware of individual housing standards, our results on the relationship of housing quality may be attenuated by our use of housing expenses rather than a measure more directly related to passing the inspection. The lower enrollment rate in St. Joseph County, particularly among renters, may be due in great part to the lower housing quality there. (See Carter and Wendt, 1982, for details of the argument.)

The remaining statistically significant variables show that owners and those who do not oppose government housing aid were more likely to enroll. Life cycle has a weak effect, with households with children enrolling more frequently and nonelderly childless couples enrolling less frequently than the elderly. Minority status was not statistically significant and was dropped from the equation.

PARTICIPATION BY ENROLLEES

An enrolled household had to be living in a dwelling that met program standards before receiving an allowance. In roughly half the cases, the dwelling at enrollment was found to be adequate at its first inspection and payments began immediately. Among those who failed the housing standard, households with less serious defects could more easily repair their dwellings and were more likely to receive payments. However, the household's response to failure also depended on other factors that affect the value of the allowance to the family. We begin this section with a demographic description of the incidence of substandard housing among program enrollees. Then we report on the factors that explain the household's response to failure on the enrollment dwelling.

Because the analysis uses data on which housing standards were not met, we restrict the data set to households whose enrollment dwelling was completely evaluated. Approximately 6% of Brown County enrollees and 12% of St. Joseph County enrollees did not complete an enrollment dwelling evaluation. Most of these terminated without receiving payments, and thus the statistics presented in this section will not match those presented in Table 5.2.

The rental housing stock is in worse condition in St. Joseph County than in Brown County, as evidenced by the greater failure rate (59% versus 45%). However, the failure rate for owners is similar in both counties (see Table 5.6). In both sites failure rates are higher for minorities than for whites and lower for the elderly than the nonelderly. These demographic characteristics are, however, interrelated. Nonwhites are more likely to be renters than whites, while the elderly are more likely to be owners than the nonelderly. In order to get a better understanding of which households fail, we modeled failure rates as a logit function of household characteristics.

The results are given in Table 5.7. As one would expect, failure rates

TABLE 5.6 Failure Rates on Initial Enrollment Evaluations

	Percentage Failing	
Household Group	*Brown County*	*St. Joseph County*
Total	47	55
Renters	45	59
Owners	51	52
Elderly	36	46
Nonelderly	51	61
White, non-Latin	46	52
Minority	64	66
Sample sizes	(8,163)	(10,417)

SOURCE: Coleman (1982: Tables 2 and C.1).

NOTE: Excludes households that did not receive a complete initial evaluation of their enrollment dwelling.

increased with the amount of allowance the household was entitled to receive. However, the magnitude of the effect was quite small. The low income elasticity of the demand for housing (see Chapters 8 and 10) means that the poorest segment of the eligibles are spending very little less for housing than other eligibles. The link between income and standard housing is further weakened by the lack of perfect correspondence between rent and the likelihood that housing meets program standards of safety and sanitation.

Minorities and larger households have much larger failure rates than their

TABLE 5.7 Logit Description of Failure Rates

Variable	*Coefficient*	*t-statistic*	*Values Compared*	*Change in Percentage Failing*
		Brown County		
Annual allowance ($000)	.110	2.02	525 to 1154[a]	1.6
Household size	.214	9.07	2 to 4	9.2
White, non-Latin	−.588	4.72	nonwhite to white	−13.6
Elderly couple	−.142	1.33	no to yes	−3.1
Elderly single	−.146	1.87	no to yes	−3.6
Single parent	.061	.81	no to yes	2.4
Others with children	.274	2.95	no to yes	7.9
Owner	.160	3.00	renter to owner	3.9
Time	−.099	3.00	Jan. 1976 to Jan. 1977	−2.3
New paint standard	.207	2.07	before to after	4.8
Constant	7.198	2.88		
(Sample size=8,163; chi-square=534)				
		St. Joseph County		
Annual allowance ($000)	.130	2.51	536 to 1293[a]	2.3
Household size	.232	10.32	2 to 4	9.2
White, non-Latin	−.319	6.13	nonwhite to white	−7.2
Elderly couple	−.309	3.61	no to yes	−7.0
Elderly single	.068	1.02	no to yes	1.0
Single parent	.142	1.94	no to yes	4.7
Others with children	−.032	−.35	no to yes	−1.4
Tenure	−.112	2.47	renter to owner	−2.5
Time	−.283	5.78	Jan. 1976 to Jan. 1977	−6.7
New paint standard	.403	4.95	before to after	9.5
First program year	−.236	3.48	no to yes	−5.6
Constant	21.145	5.72		
(Sample size=10,147; chi-square=593)				

SOURCE: Coleman (1982: Tables 4 and C.2).
a. Values compared are the 25th to the 75th percentile.

opposite groups. The St. Joseph County minority effect is a better indication of the disparity between minorities and whites than the Brown County effect. (Brown County has very few nonwhites.) After controlling for household size, race, and allowance entitlement, the effects of tenure and life cycle are much smaller than they appear in the uncontrolled data of Table 5.6.

The remaining variables in the model show how failure rate changed over time. Beginning in January 1977, households with small children were not allowed to receive payments if flaking or peeling paint was present in their dwelling. The new paint standard increased the failure rate by 5% in Brown County and almost 10% in St. Joseph County. Controlling for the effect of the paint standard, failure rates declined over time. This might be because some housing repaired by previous recipients was being evaluated for new enrollees or because more widespread knowledge of the housing standards inhibited more of those in substandard housing from enrolling. However, if self-selection was the primary reason for the decline in failure rates, one would have also expected to find a corresponding decline in the termination rate following failure; this did not occur.

Renter response to failure of the housing standards was modeled using discriminant analysis on the three possible outcomes–moving, repairing, or terminating. Since less than 3% of owners moved following dwelling failure, we used a logit equation with the binary dependent variable being whether or not the household terminated from the program without receiving payments.

Separate equations were fit for each site and tenure group. Table 5.8 presents results for St. Joseph County renters. The discussion below will point out important difference for other groups.[12] In general, households responded to both the costs and benefits of the allowance program in deciding whether to obtain an allowance. In all four site and tenure combinations, those with larger allowance entitlements were less likely to terminate than similarly situated households with smaller entitlements. Program costs were measured with dummy variables for the types of housing defects, the number of defects, and the average dollar cost to repair dwellings with a similar set of defects. The latter variable measures only out-of-pocket expenses and does not include the value of unpaid labor. In general, those who did repair their dwelling did so very inexpensively; St. Joseph County renters averaged $49, Brown County renters, only $29.

Two of the most important deterrents to obtaining an allowance were failures of the occupancy[13] and paint standards. However, failure of the occupancy standard also led many renter households to move to a more suitable dwelling. Among St. Joseph County renters, failure of the paint standard led only to increased terminations, not moves. In the other three site and tenure combinations, the paint standard also significantly increased terminations beyond what would be estimated from the dollar cost to repaint the affected area.

In St. Joseph County, renters in larger multiple dwellings were less likely to repair than those in smaller buildings. It may be that certain types of re-

TABLE 5.8 Results of Discriminant Analysis of Response to Failure on Initial Housing Evaluation for St. Joseph County Renters

		Changes in Percentage Responding		
Factor	*Values Compared*	*Move*	*Repair*	*Terminate*
Allowance*	\$668 to \$1623/yr.[a]	1.9	11.4	−13.3
Duration of eligibility*,[b]	1.9 to 9.6 yrs.[a]	1.0	2.0	−3.0
Estimated repair costs*	\$14.19 to \$52.14[a]	0.6	−4.2	3.6
Number of defects	1-3	2.0	1.5	−3.5
Occupancy standard*	pass to fail	14.6	−29.5	14.9
Paint standard*	pass to fail	−3.1	−20.8	23.9
Stairs, railings*	no to yes	−0.6	−4.6	5.2
Plumbing, wiring	no to yes	0.2	1.9	−2.1
Other hazards*	no to yes	4.3	−5.8	1.5
Other defects*	no to yes	1.0	−13.1	12.1
Single family residence*	no to yes	−5.3	6.9	−1.6
Duplex	no to yes	−4.5	9.6	−5.2
Row-house*	no to yes	0.1	9.8	−9.9
Length of stay	0.41 to 2.08 yrs.[a]	−0.1	0.3	−0.2
Reduced Rent	no to yes	−1.1	−1.6	2.7
Race	nonwhite to white	−0.9	1.3	−0.4
Household size*	2 to 4	−3.6	0.0	3.6
Elderly couple	no to yes	4.5	2.4	−6.9
Elderly single	no to yes	−7.3	16.1	−8.8
Single parent	no to yes	5.8	−0.2	−5.6
Others with children	no to yes	3.5	−7.1	3.7
Time	Jan. 1976 to Jan. 1977	−2.9	4.0	−1.1
First program year*	no to yes	8.8	−8.9	0.1
Sample size=2,422				

SOURCE: Coleman (1982: Tables 8 and 9).
*Variable statistically significant at 5% level or better.
a. The values compared are the 25th to the 75th percentile.
b. Entered in equation in log form.

pairs are more difficult in multiple unit dwellings because the tenant has less control over the entire property. Although the direction and magnitude of the effect was similar in Brown County, it was not statistically significant there (see Coleman, 1981: Table 7).

Elderly singles were the group most willing to repair their dwellings. Household size had a smaller positive effect on termination rates. The other demographic effects were small and varied by site and tenure.

OUTCOME OF THE PARTICIPATION PROCESS

Participation decisions result from balancing the benefits of participation with the costs of obtaining adequate housing. However, the costs of participation are larger for the poorest households, so it is important to know whether the program reaches more of the poorest households despite their poorer initial housing quality. We also consider how participation varies with

TABLE 5.9 Logit Model for Probability of Receiving Payments at Fourth Survey Wave

Variable	*Coefficient*	*t-statistic*	*Values Compared*	*Change in Percentage Receiving Payments*
Annual benefit level (1n)	.396	3.70	508 to 1275	8
Duration of eligibility	.036	5.18	1 to 14	10
Fraction of previous year eligible (1n)	1.143	3.01	.7 to 1.0	10
Household size	−0.206	3.27	2 to 4	−9
Child in household[a]	1.359	3.68	no to yes[b]	28
Single parent[a]	−0.388	1.70	no to yes[b]	18
Elderly[a]	0.089	0.24	no to yes[b]	0
Elderly couple[a]	−0.196	0.70	no to yes[b]	−3
Owner	0.067	0.35	no to yes	1
Minority	0.164	0.65	no to yes	4
St. Joseph County	−0.456	2.88	no to yes	−10
Previous interview	0.571	3.35	no to yes	13
Residual length of stay (months)	−0.004	2.78	−24 to 15	−3
Interaction of residual and owner	.005	2.79	−24 to 15	+3
Constant	−3.281	3.70		

(Sample size=973; chi-square=165.6)

SOURCE: Carter and Wendt (1982: Tables 6.3 and 6.4).
a. Set of four life cycle variables yield a chi-square of 29.9 with 4 degrees of freedom.
b. Comparison is from a childless nonelderly household to the indicated group.

other household characteristics that either lead to difficulty in obtaining adequate housing (e.g., larger households) or that could be related to attitudinal differences toward welfare or to the difficulty in acquiring knowledge (e.g., the elderly). We use a multivariate logit model to answer the question of whether there are identifiable groups of households that participate less than others at the same level of need.

The data are from the fourth household survey wave in each site. The dependent variable is the probability that an eligible household was receiving an allowance on the date of survey. Having been interviewed at an earlier survey wave increases the probability, and consequently a dummy variable for previous interviews is used as a control. Since the survey was addressed to the same housing unit each year, there is a high correlation between previous interview and mobility. To remove the effects of household mobility from all coefficients, we include the residual from a regression of length of stay in the surveyed unit on all other independent variables in the model (except previous interview)[14].

The results are given in Table 5.9. The display of changes in the probability of being a recipient was calculated with interview status set to 0 (no previous interview) and all other variables held at their mean value.

The variables used to measure need are allowance entitlement and duration of eligibility. Although poor housing does reduce the participation rate of those with large benefits slightly more than those with smaller benefits, households with larger benefits are still more likely to be in the program.

As discussed previously, duration of eligibility affects participation in two ways: (1) A longer expected duration of eligibility motivates households to apply for allowance and to comply with the housing standard, and (2) the pool of eligibles always contains some households that are newly eligible and have not yet applied for an allowance, although they will if they remain eligible long enough. These two effects are best modeled by including two variables: the expected duration of eligibility and the log of the expected fraction of the previous year which the household spent eligible.

As the duration of eligibility varies from 1 to 14 years (the twenty-fifth to seventy-fifth percentile in the data) and the expected fraction of previous year eligible varies from 0.7 to 1.0, the total effect on participation rate is an increase from 23% to 43%. Since the effects on participation of higher benefit levels and longer expected duration of eligibility are positive and very significant, we conclude that despite housing standards, the program is serving the neediest households more frequently than the less needy.[15]

As household size increases, participation in the program decreases. The difficulty larger households face in participating may be traced directly to their problems securing standard housing. Enrollees from larger households are much more likely to fail the housing evaluations and more likely to fail the occupancy standard (which is the most difficult to remedy). In addition, the effect of household size on the enrollment decision is trivial, indicating that the decrease in participation observed here is due to the difficulty larger households have in obtaining adequate housing.

Five household types are distinguished in this analysis: married couples with children, single parents, elderly couples, elderly singles, and the default group of married couples without children. Controlling for need, households with children participate more than the other groups. No other statistically significant distinctions among groups can be found in the data. There also are no statistically significant effects of tenure or race.

Benefit level, duration of eligibility, and household size are major determinants of participation, and these vary in a systematic way by life cycle, tenure, and race. This variation leads to actual participation patterns that are quite different from the marginal effects of household characteristics just discussed. The life cycle groups with the largest participation rates are actually elderly singles and single parents (see Carter and Balch, 1981). Couples with children participate less than these groups because of their larger household size and their shorter expected duration of eligibility. Renters participate much more than owners, but this is completely due to other variables in the model.

Households in St. Joseph County participate at a rate 10 percentage points lower than similar Brown County households. No interactions between site and other explanatory variables could be found in the data. We believe that the lower participation in St. Joseph County is due in great part to the lower quality of the housing stock.

CONCLUSIONS

Because HASE operated an open enrollment program, our participation rates were influenced by two factors that could not be considered in HADE: program knowledge and delays during the participation process. The vigorous program publicity campaigns conducted during the early program years informed 85% of the eligible population about the program. Even hard-to-reach groups, such as the elderly, were successfully informed through special targeted outreach programs in St. Joseph County. Using the behavioral equations, we estimate that steady state participation rates would increase by only 6% of the eligible population if all households had been informed.

It takes about 6 months for a household to enroll and an additional 1.5 months to complete the housing certification process. In addition, other households will never enroll, only because they become ineligible before they get around to enrolling. We estimate that all these delays account for 10% of the eligible population being nonparticipants at any point in time.

Both HADE and HASE data confirm the importance of the housing standards in deterring participation. In HASE, households in better housing were more likely to make it through each step of the participation process. The largest effects of housing were found after enrollment: Enrollees whose dwellings passed the standards immediately received payments; the cost of repair, number of defects, and type of defects were important predictors of the subsequent participation of enrollees whose dwellings failed. In Carter and Wendt (1982) we estimated that if housing standards were eliminated, steady state participation rates would increase by 14% of the eligible population in Brown County and 25% in St. Joseph County.

The findings from HASE data differ from those of HADE in one important respect. In HASE, program benefits were one of the major determinants of participation and exerted a positive influence at each step in the process. Increases in program benefits—the dollar amount of the allowance to which the household was entitled and the length of time they could have expected to receive it—spurred households to join the program and to comply with housing requirements. Although the probability that a household was already living in standard housing at enrollment declined slightly with increasing benefits, the greater response of the poorer enrollees whose dwellings failed more than offset this difference so that poorer enrollees were still more likely to receive a housing allowance.

The reported HASE and HADE participation rates are not directly comparable. HASE rates refer to the percentage of eligibles receiving payments at a point in time; HADE rates refer to the percentage of those invited to join who ever qualified for payments. Thus the HADE rates eliminate the effects of both knowledge and participation delays. However, the percentage of knowledgeable eligible households in the HASE sites that ever received payments can be estimated from Table 5.2. This first approximation to participation in an invitational enrollment program yields renter participation rates of 65% in Brown County and 54% in St. Joseph County.[16] These figures are higher than the HADE rate of 45% in Phoenix and 30% in Pittsburgh. At least part of the remaining cross-experimental difference is undoubtedly due to the differing housing standards used.

NOTES

1. A fuller account of the research summarized here can be found in Carter and Wendt (1982), Coleman (1982), Wendt (1982), Ellickson (1981), and Carter and Balch (1981).

2. Excluding nonelderly singles, most of whom were first able to join in August 1977.

3. The fourth survey wave in Brown County was conducted during the first six months of 1977, and the third year of program operation ended in June 1977. The St. Joseph survey was one year later, and the third-year program operations ended December 1977.

4. Coleman (1981) shows that only 3% of recipients terminate each year after failing their annual housing inspection.

5. Studies showing that respondents can provide opinions about fictitious programs (see Ellickson and Kanouse, 1979) led us to suspect that many of those who could not provide details about the program did not, in fact, know about it. Additional analysis by Wendt (1981) is consistent with this hypothesis.

6. Only one record was used for each household. A random procedure was used to select the record (i.e., wave) used in the model for households interviewed more than once.

7. Housing expenses were measured by monthly rent for renters, and cash outlay for owners consisting of mortgage payments, taxes, insurance, utilities, and repairs.

8. Households that did not know of the program were not questioned any further about it.

9. Households could name more than one "most important source" of information.

10. A random procedure was used to select the record for households that met this criterion at both waves.

11. Measured as persons per bedroom.

12. See Coleman (1981) for a full discussion, including quantitative results for Brown County renters and owners in each site.

13. The occupancy standard was based on household size and number of habitable rooms. In some cases a room could be made into a habitable room without major construction.

14. See Carter and Wendt (1982), Section VI, for further justification of this procedure and supporting data.

15. To be precise, this analysis shows only that within the groups of households that are controlled for in the regression, a greater percentage of the needier households are being served

than of the less needy. However, when the dummy variables were omitted from the regression, we obtained the same result.

16. See Carter and Wendt (1982) for details of a closer approximation (69% in Brown County and 56% in St. Joseph County).

6

Participation Under Random Assignment

STEPHEN D. KENNEDY and JEAN E. MacMILLAN

☐ THE CENTRAL FINDING of the analysis of participation in the Demand Experiment is that the imposition of housing requirements sharply reduces participation among households that would not normally live in required housing in the absence of the allowance program. This has important implications for the potential size of an allowance program, the demographic composition of program recipients, and the nature of program impacts. At the same time, the analysis raises some puzzling questions of interpretation which will require more exact behavioral specifications to resolve.

ANALYTIC MODEL

The analysis of participation starts with a reasonably simple model—that a household will choose to participate in a housing allowance program if the program payments are large enough to compensate the household for the monetary and nonmonetary costs of participating. Thus, for example, a household in the percent of rent or unconstrained plans must overcome any reluctance it may have to accepting money from the government and must comply with a variety of reporting, interview, and housing inspection requirements. Housing gap households faced the same general costs to participation but also had to meet housing requirements. For these households, the net value of the allowance offer must be not only positive, but great enough to overcome the costs of obtaining standard housing.

Our analysis of participation focused on the differences in participation introduced by the housing requirements. The initial analysis (Kennedy et al., 1977) approached this problem through what seemed to be a promising and

---

AUTHORS' NOTE: *The material in this chapter is based almost entirely on analyses by Stephen D. Kennedy, T. Krishna Kumar, Jean MacMillan, Steven Sicklick, Michael Murray, and Glen Weisbrod, reported in Kennedy et al. (1977) and Kennedy and MacMillan (1980).*

relatively straightforward application of standard economic demand theory. The payment needed to compensate a household for meeting requirements is obviously zero if the household would normally meet requirements by itself. For households with less than required housing, it is the additional cost involved in obtaining required housing minus the value (to the household) of the additional housing. As it turns out, the value of additional housing can be expressed in terms of the household's demand function for housing—the schedule of amounts it will be willing to spend for housing under different housing prices and income levels. Thus analysis of participation in these terms promised results that would be both readily understandable and clearly linked to widely available information on normal housing expenditures.[1]

Unfortunately, the model failed. First, it turns out that incorporating general costs of participation, apart from those involved in meeting housing requirements, is not straightforward and can materially obscure interpretation of estimated forms.[2] More important, the model is essentially static. It comes from models in which goods are immediately available at known prices. But we know that households must search for housing, that they generally cannot specify their exact housing bundle in advance, and that the price of housing (the cost of comparable units) varies.[3]

Indeed, preliminary analysis of participation model in Kennedy et al. (1977: chap. 4) immediately suggested strong effects from either a resistance to participation in any transfer program or the importance of dynamic considerations involving search and moving behavior. The importance of moving was directly suggested by the fact that variables associated with a household's normal propensity to move had strong positive effects on the probability of participation among housing gap households. Further, participation rates for households were lower than would be expected, given the apparent costs involved in meeting requirements; indeed, the estimated coefficients indicated that an increase in minimum rent requirements of $1 would require increases in payments of from $1.5 (Pittsburgh) to $2.5 to maintain participation rates at a given level. This again suggested either substantial costs involved in moving to meet requirements or substantial misestimation of the rent levels associated with meeting requirements. In addition, analysis of the initial enrollment decision (apart from subsequently meeting requirements) indicated a significant participation loss unrelated to housing requirements.

Given the difficulties involved in developing and empirically identifying the parameters of models that take explicit account of such dynamics, we decided instead to rely on the availability of control observations to develop specifications based on a loosely constructed specification of the propensity to occupy standard housing. Consider first a household that originally lives in housing that does not meet program requirements. It may come to meet requirements either by upgrading its current unit or by moving to a unit that meets requirements. The probability of either of these events would be expected to depend on the distance (somehow defined) between the household's original housing and required housing, and on the probability that the household moves. Further, the relationship between initial and future distance would be expected to be stronger if the household did not move; thus, the probability of moving and distance would be expected to interact.

Now consider the effect of the allowance offer on such households. This will clearly depend on the size of the allowance payment offered, but may also depend on the household's resistance to moving, especially as its distance from meeting requirements increases (and hence, its ability to "upgrade" its current unit decreases). In short, the probability of participation, P, is roughly specified as a function of some measure of distance from meeting requirements (D), the household's normal probability of moving (P_m), the allowance payment (S), and the "cost" of moving for households that would not normally move (C_m). Lacking any strong notions about the determinants of any general resistance to participation apart from the housing requirements, we simply added a variety of plausible and/or interesting demographic descriptors, and wrote

$$P = K(D, P_m, S, C_m, X) \quad [6.1]$$

where

K = some distribution function;

X = a vector of demographic descriptors.

For convenience, the distribution function was assumed to be logistic.

The specific demographic descriptors used included various dummy variables for the age and race/ethnicity of the household head and for household size and composition, expressed satisfaction with housing, participation in other transfer programs, and income. The four behavioral descriptors varied in details depending on the analysis involved, but were essentially defined as follows:

Distance (D) was defined so that large negative values indicate greater distance—that is, as the difference between the household's enrollment housing expenditures (R_o) and an estimate of the cost of required housing (R_R):

$$D = R_o - R_R \quad [6.2]$$

Enrollment rents may, of course, deviate from "normal" rents. However, there is a strong serial correlation in rental expenditures, even among households that moved, so that the error is not great.[4] The cost of required housing was estimated by the actual minimum rent requirement for minimum rent households and by the estimated cost of standard housing used in the payment formula for minimum standards households. This was intended to take account of the differences in requirements associated with differences in household size. It should be noted, as discussed later, that these variables probably underestimate the actual cost of obtaining minimum rent housing.

Probability of moving. For the analysis of participation after enrollment, this variable was constructed from a logistic estimation of normal moving behavior based on observation of control households (MacMillan, 1980). The analysis of acceptance did not include this variable, relying instead on a prior mobility variable (number of moves in the three years prior to the experiment).

Allowance payment. In the analysis of acceptance, the allowance payment variable was simply the preliminary payment estimate given to households when they were offered enrollment.

The analysis of participation after enrollment was based on the payment the household would receive if it met all requirements, as calculated by the program office. In order to separate payment effects from demographic correlates, the payment variable was further broken into two parts—a payment level variable defined as the payment that the household would receive at the modal household size (4) and average income of all enrolled households, and a residual payment equal to the difference between the payment level and the household's actual potential payment. The payment level variable was a function of treatment parameters alone, while the residual payment variable reflected both treatment variations and household characteristics.

Cost of moving. This variable is the expected cost to the household of moving to meet requirements. It was used only in the analysis of participation after enrollment and was defined as a transform of the estimated probability that the household would move normally.[5]

It should be noted that while these variables are plausible enough, the specification itself is not derived from any clearly stated behavioral model. This is, of course, quite common in empirical analysis. The tremendous advantage of experimental situations (essentially the availability of control observations and treatment variations) is that it allows strong interpretation of weak models.

EMPIRICAL RESULTS

OVERALL PARTICIPATION RATES

Participation in the Demand Experiment involved a number of stages—being contacted for enrollment, completing the enrollment interview, deciding to accept the enrollment offer, being determined eligible, enrolling, and becoming a recipient. For the present analysis, these stages were combined into two major participation decisions—first, accepting the enrollment offer and enrolling in the experiment and, second, once enrolled, actually participating in the program and receiving an allowance payment. The analysis of acceptance is based on households that got far enough in the enrollment process to receive a complete description of the program offer. The analysis of subsequent participation is based on enrolled households. Participants are defined as all enrolled households that *ever* received an allowance payment over the two years of the experiment. Overall participation rates are the product of the acceptance rate and the subsequent participation rate.

All households had to decide whether or not to accept the enrollment offer. Once enrolled, eligible percent of rent and unconstrained households began to receive payments immediately. Housing gap households, on the other hand, had to meet the housing requirements of the program in order to participate. Thus, the second stage in the participation process is a direct result of the imposition of housing requirements. In theory, households could, of course, anticipate the effect of housing requirements in deciding whether to accept enrollment. In fact, the effect of housing requirements was largely confined to the second stage.

Table 6.1 shows the overall participation, acceptance, and subsequent participation rates for each of the three program types, as well as a breakdown for the housing gap subprograms defined by the different housing requirements. Overall participation rates were generally somewhat higher in Phoenix than in Pittsburgh. Within each site, they are very similar and reasonably high (about 84%) for percent of rent and unconstrained households, the two program types that did not impose housing requirements. They are much lower for the housing gap plans—45% or roughly half as large as for the programs without any housing requirements. There is considerable variation in participation rates among different housing requirements used in the housing gap plans; the minimum rent low plans have rates about half again as large as the rates for minimum standards and minimum rent high. Most of the differences in overall participation arise from differences in subsequent participation after enrollment rather than differences in acceptance of the enrollment offer.

ACCEPTANCE

As indicated in Table 6.1, acceptance rates did differ among the three programs and across the housing gap subprograms. The differences are, however, relatively minor compared to the difference in overall participation. Indeed, in many ways, the key finding in the analysis of acceptance is what does not matter. There are no consistent demographic differences and only small differences among programs. Once payments offered reach a level of $40 a month, acceptance rates for most programs will apparently reach 80%-90%. The remaining 10%-20% will refuse the program for a variety of reasons, most commonly having to do with the bother of program requirements and reluctance to accept money from the government.

Objection to program requirements was the reason most frequently (about half of the households) given for turning down the enrollment offer. Objections to requirements included objections to the housing requirements im-

TABLE 6.1 Stages in Participation (percentages; sample sizes in parentheses)

	Combined Sites			*Pittsburgh*			*Phoenix*		
	Acceptance Rate	*Subsequent Participation Rate*	*Overall Participation Rate*[a]	*Acceptance Rate*	*Subsequent Participation Rate*	*Overall Participation Rate*[a]	*Acceptance Rate*	*Subsequent Participation Rate*	*Overall Participation Rate*[a]
Percentage of rent households	84 (1499)	100 (960)	84	82 (821)	100 (484)	82	87 (678)	100 (476)	87
Unconstrained households	83 (209)	100 (143)	83	78 (120)	100 (73)	78	90 (89)	100 (70)	90
Control households	69 (1613)	100 (952)	69	61 (863)	100 (431)	61	78 (750)	100 (521)	78
Housing gap households	78 (2093)	58 (1254)	45	74 (1086)	56 (592)	41	83 (1007)	59 (662)	49
Minimum standards requirement	79 (959)	47 (575)	38	75 (489)	40 (268)	30	84 (470)	54 (307)	45
Minimum rent low requirement	78 (545)	77 (323)	60	74 (287)	81 (156)	60	82 (258)	74 (167)	61
Minimum rent high requirement	77 (589)	56 (356)	43	73 (310)	58 (168)	42	81 (279)	54 (188)	44

SOURCE: Kennedy and MacMillan (1980: Tables 2.3 and 2.4).

a. Defined as the product of the acceptance and subsequent participation rates.

TABLE 6.2 Logit Estimation of the Probability of Accepting the Enrollment Offer (sample sizes in parentheses)

	Pittsburgh			*Phoenix*		
	Coefficient	*t-statistic*	*Partial Derivative*[a]	*Coefficient*	*t-statistic*	*Partial Derivative*[a]
Constant	−1.310	−2.80**	NA	−0.024	−0.08	NA
Elderly household	−0.280	−1.69†	−0.048	0.153	0.70	0.020
Young household	0.369	2.15*	0.064	0.138	0.82	0.018
Black household	0.420	3.17**	0.069	−0.778	−3.02**	−0.101
Spanish-American household	NA	NA	NA	−0.663	−4.10**	−0.086
Large household	0.476	3.20**	0.082	−0.006	−0.03	−0.001
Single-parent household	0.144	0.97	0.025	0.298	1.68†	0.039
Prior mobility	0.248	4.06**	0.043	0.168	3.55**	0.022
Dissatisfaction with unit or neighborhood	0.248	2.63**	0.043	0.138	1.08	0.018
Participation in other transfer programs	0.085	0.68	0.015	0.135	0.95	0.018
Income (in thousands)						
Under $3000	0.335	2.19*	0.060	−0.076	−5.69**	−0.010
$3000-$8000[b]	−0.122	2.87**	−0.021	−0.003	−0.08	0.000
Over $8000[b]	−0.058	−0.21	−0.010	−0.069	−1.17	−0.089
Estimated subsidy amount						
Under $40	0.042	6.52**	0.007	0.051	5.72**	0.007
$40-$80	0.018	2.87**	0.003	0.008	1.17	0.001
Over $80[b]	−0.009	−1.84	−0.002	0.004	0.98	0.002
Unconstrained household	0.011	0.05	0.002	0.472	1.49	0.061
Percent of rent household	0.506	4.31**	0.087	0.343	2.45*	0.045

SOURCE: Kennedy and MacMillan (1980: Table 3.3).

a. The partial derivative with respect to the ith variable is calculated as

$$\frac{dp}{dx_i} = p(1-p)\,b_i$$

where p is the mean probability and b is the estimated logistic coefficient for the ith variable.

b. These were estimated as splines. The coefficients shown in the table are the estimated coefficients applicable in the range stated.

†t-statistic significant at the 0.10 level (two-tailed).

*t-statistic significant at the 0.05 level (two-tailed).

**t-statistic significant at the 0.01 level (two-tailed).

posed on housing gap households, but they were apparently predominantly concerned with various reporting requirements (including monthly income reports and periodic interviews, as well as regular housing inspections and submission of rent receipts). Objection to participation in a government program was second in frequency. Over 40% of the households at both sites mentioned that they did not enroll because they did not want to accept charity or otherwise objected to the idea of accepting money from the government. A number of other reasons were cited, but none accounts for more than about a quarter of the households that rejected the offer at either site.

Overall, the same considerations seem to have influenced acceptance in both sites. The bother and paperwork of participating and general objections to accepting money from government programs were mentioned most frequently, but most households cited some other reason as well. The only statistically significant differences were more frequent expressions of concern about eligibility and payment amounts in Phoenix.[6] On the other hand, the relation between acceptance and demographic characteristics, while not strong, was quite different in the two sites. Table 6.2 presents the estimated coefficients for the logistic specification of equation 6.1. Homogeneity tests consistently showed a significant loss of explanatory power when demographic effects in the two sites were constrained to be the same. Further, this was true regardless of whether equations were estimated for all treatment groups, as in Table 6.2, or for each major group separately.

Thus, for example, the largest difference in acceptance rates across demographic groups in either site was that associated with age in Pittsburgh. The acceptance rate among elderly households in Pittsburgh was 61%, 20 points lower than the 81% rate for households where the head of household was under 30. Taking account of other demographic factors associated with age, however, the difference in acceptance rates between elderly and young households, shown in Table 6.2, was 11 percentage points. In Phoenix, on the other hand, the difference was only 7 points (4 percentage points taking account of other demographic factors) and not statistically significant. Other differences among demographic groups were smaller but also inconsistent across the two sites. Black households accepted enrollment somewhat more often than whites in Pittsburgh, but both blacks and Spanish Americans were less likely to accept in Phoenix. The poorest households (those with incomes of less than $2000) accepted less often in Pittsburgh but not Phoenix. Welfare recipients were more likely to accept in Pittsburgh but not Phoenix, and when other demographic factors were taken into account, welfare status had no significant effect in either site.

Overall, then, it appears that program acceptance will vary across demographic groups but that it will vary differently from place to place. However, demographic differences in acceptance were not significantly related to differences in program type.[7] Thus, it appears that these differences, when they arise, might be expected to apply equally to all programs in an area.

The only consistent patterns found in both sites were a positive relation between acceptance rates and allowance payment and the somewhat lower acceptance rates for housing gap as compared to percent of rent households. Between payments of roughly $10 and $40 per month, higher payment estimates were associated with sharply higher acceptance rates. Average acceptance rates in the two sites rose from 60% to 67% for households with $10 payment estimates to over 85% for those with estimates of about $40 per month. Thereafter, acceptance rates are almost level. Acceptance rates were also significantly, though modestly, lower for housing gap, as compared with percent of rent. Further analysis showed, however, that housing gap acceptance rates were lower only among higher income households with incomes greater than 80% of the eligibility limits. It appears that higher income households have a more negative reaction to the idea of housing requirements, regardless of the payment offered them.

Once the payment amount estimated at enrollment was taken into account, there was no evidence of significant differences in acceptance associated with variations in the percent of rent or housing gap payment formulas. Likewise, there is no evidence that households' actual probability of meeting the requirement or their perceptions about whether they already met the requirements had any effect on their propensity to accept enrollment.

SUBSEQUENT PARTICIPATION OF ENROLLED HOUSEHOLDS

After households had enrolled, housing requirements played a critical and highly specific role in participation. For percent of rent households, unconstrained households, and housing gap households that already met housing requirements when they enrolled, the decision to enroll was the participation decision. Once these households accepted the enrollment offer and were certified as eligible, they began to receive allowance payments immediately. For housing gap households that did not already meet housing requirements, however, program participation involved another step. In order to participate, these households either had to arrange to meet requirements in their enrollment unit or move to a different unit that did meet the requirements. It is this additional step that accounted for most of the difference in overall participation rates shown in Table 6.1.

The subsequent participation rate of housing gap households after enrollment depends on two factors—the proportion of households that already met requirements when they enrolled (all of which participated immediately) and the willingness of households not already in acceptable housing to change their housing in order to meet requirements and participate. Table 6.3 shows how the subsequent participation rate for each type of requirement was determined by these two factors. Requirements that had the highest proportion of households already meeting them at enrollment also tended to have higher participation rates among households that did not meet the requirements at enrollment. There were also substantial differences in subsequent participation rates by demographic groups. As it turns out, however, all of these demographic differences derive from differences in allowance payments and in the normal propensity to occupy required housing.

TABLE 6.3 Initial Payment Status and Subsequent Participation (percentages; sample sizes in parentheses)

	Pittsburgh				*Phoenix*			
	All Housing Gap Households	*Minimum Standards Requirement*	*Minimum Rent Low Requirement*	*Minimum Rent High Requirement*	*All Housing Gap Households*	*Minimum Standards Requirement*	*Minimum Rent Low Requirement*	*Minimum Rent High Requirement*
Percentage of enrolled households that received a full payment at enrollment	33 (592)	15 (268)	64 (156)	35 (168)	29 (662)	19 (307)	53 (167)	27 (188)
Subsequent participation rate for households that received a full payment at enrollment	100 (197)	100 (39)	100 (100)	100 (58)	100 (195)	100 (57)	100 (88)	100 (50)
Subsequent participation rate for households that did not receive a full payment at enrollment	34 (395)	30 (229)	48 (56)	35 (110)	42 (467)	44 (250)	46 (79)	37 (138)
Subsequent participation rate for all enrolled households	56 (592)	40 (268)	81 (156)	58 (168)	59 (662)	54 (307)	74 (167)	54 (188)
Percentage of all participants that received a full payment at enrollment	60 (331)	36 (107)	79 (127)	60 (97)	50 (391)	34 (166)	71 (124)	50 (101)

SOURCE: Kennedy and MacMillan (1980: Table 2.9).

TABLE 6.4 Comparison of the Participation Rate for Housing Gap Households with the Rate at which Control Households Met Requirements— Combined Sites

	Minimum Standards Requirement			*Minimum Rent Low Requirement*			*Minimum Rent High Requirement*		
	Coefficient	*t-statistic*	*Partial Derivative*[a]	*Coefficient*	*t-statistic*	*Partial Derivative*[a]	*Coefficient*	*t-statistic*	*Partial Derivative*[a]
Constant	−1.303	−5.82**	NA	−0.935	−2.59**	NA	−0.607	2.15*	NA
Distance (units of $10)	−0.172	−5.12**	−0.034	−0.485	3.54**	−0.104	−0.506	5.62**	−0.090
Probability of moving (units of .10)	0.069	1.60	0.014	0.234	3.82**	0.050	0.178	3.43**	0.032
Phoenix households	0.563	1.77†	0.112	−0.637	−1.24	−0.137	−0.652	−1.47	−0.116
Distance in Phoenix (units of $10)	0.056	1.47	0.011	0.342	2.30*	0.073	0.298	3.04**	0.053
Probability of moving in Phoenix (units of .10)	−0.011	−0.21	−0.002	−0.081	1.06	−0.017	−0.059	−0.87	−0.010
Housing gap households	−0.329	−1.04	−0.066	0.585	0.86	0.125	0.240	0.54	0.043
Payment level (units of $10)	0.186	4.89**	0.037	0.051	0.56	0.011	0.093	1.53	0.017
Residual payment (units of $10)	0.036	0.95	0.007	−0.016	−0.20	−0.003	0.080	1.41	0.014
Likelihood ratio (significance)		125.584**			57.050**			107.127**	
Sample size		1046			478			797	
Mean of dependent variable		0.275			0.312			0.231	
Coefficient of determination		0.102			0.096			0.124	

SOURCE: Kennedy and MacMillan (1980: Table 4.4).
a. Derivatives computed at sample mean.
†t-statistic significant at the 0.10 level (two-tailed).
*t-statistic significant at the 0.05 level (two-tailed).
**t-statistic significant at the 0.01 level (two-tailed).

The analysis of subsequent participation started with the logistic specification of equation 6.1. Three basic equations were involved, one for each of the three housing requirements. In each case, the equation was estimated using control households (to provide data on the normal probability of meeting requirements) and the housing gap households subject to the requirement involved. The dependent variable involved was whether a household ever met requirements and received a full payment. Accordingly, the analysis focused on households that were not already living in program required housing when they enrolled.

The results of the analysis are presented in Table 6.4. Preliminary runs indicated that distance from meeting requirements could be adequately characterized (in terms of expenditures) by the difference between required expenditure levels and the household's enrollment expenditures. In addition, terms in the log of the probability of moving (intended to capture the effects of having to move to meet requirements) were generally insignificant and were dropped from the equation.[8] Most important, it turned out that all the demographic variables could be dropped; they yielded no significant improvement once distance, the probability of moving, and allowance payments were taken into account.[9]

As indicated in Table 6.4 there were site differences in control households' normal propensity to meet requirements as a function of initial distance and the estimated probability of moving. With one exception, however, there was no different effect of these variables for experimental households.[10] In addition, given the underlying differences in the normal propensity to meet requirements, the two sites could be pooled in each case.[11] Furthermore, despite the apparent differences in estimated experimental coefficients in Table 6.4, it appears that the allowance effects may be the same for all three requirements. Thus, for example, estimation with a single dummy variable for experimental households yields almost identical coefficients, as shown in the first row of Table 6.5.[12] Likewise, maximum likelihood estimates of common coefficients for the three requirements indicate that they can be pooled.

In sum, for households that do not already meet a given set of housing requirements when they enroll, it appears that the probability of participation is a simple function of the household's normal probability of meeting requirements and the amount of the allowance payment offered. This is a pleasantly compact conclusion. There are some clear deficiencies, however. One problem is the insignificant coefficient for the residual payment variable, which represents differences in payment associated with income and household size, as well as treatment parameters. We would have expected this variable to have the same coefficient as the payment level variable. In addition, evidence from the Supply Experiment suggests an important omission in the specification. As shown in Table 6.6, among households enrolled in the Supply Experiment, those that did not already meet requirements when they enrolled had much higher participation rates than minimum standards households in the Demand Experiment. The difference, however, is entirely accounted for by a difference in the proportion upgrading their enrollment units to meet requirements.[13] This suggests that we should have considered not simply distance from meeting requirements but also some measure of the capacity to meet requirements in place. It should also be pointed out that this

TABLE 6.5 Pooled Experimental Effects

	Minimum Standards	*Minimum Rent Low*	*Minimum Rent High*
Estimated overall experimental effect[a]	0.875	0.883	.925
(standard error)	(0.136)	(0.224)	(0.171)
	Coefficient	*Standard Deviation*	*t-statistic*
Pooled estimates[b]			
Housing gap households	−0.08502	.2383	−0.36
Payment level ($10 units)	0.1492	.003019	4.94**
Residual payment ($10 units)	0.04106	.002914	1.41
Likelihood ratio[c] statistic for pooled versus separate experimental coefficients (dof)		4.7477 (6)	

a. Estimated with a single dummy variable for experimental households replacing the three experimental variables in Table 6.5 (see Kennedy and MacMillan, 1980: Table 4.5).

b. These estimates were derived from the estimates of Table 6.5 under the assumption that each set of coefficients represented an independent draw from a population with unknown true mean, m_i, and known covariance matrix, S_i, where S_i is the estimated asymptotic covariance matrix from the logistic estimations of Table 6.5.

c. Test level is $\chi^2_{\sigma}(.10) = 10.645$.

TABLE 6.6 Comparison of Demand and Supply Experiment Participation Rates Among Households That Did Not Already Meet Minimum Standards at Enrollment

	Demand Experiment[a] Pittsburgh/Phoenix	*Supply Experiment[b] Brown County/St. Joseph County*
Percentage that first met requirements		
By upgrading enrollment unit	17	40
By moving	20	20
Never met	64	40
(Sample size)	(471)	(2597)

SOURCE: Demand Experiment—Tabulations of baseline and periodic household surveys; Supply Experiment—Rand Corporation (1978: 65).

a. Minimum standards households only.

b. Renters only.

does not contradict the earlier finding that moving *per se* was not a barrier to participation. This was based on the fact that the probability of moving affected only the normal propensity to meet requirements and had no further effect on experimental participation. This only means that reluctance to move is not in itself a barrier; it does not mean that finding a unit that meets requirements to which to move is no problem.

Finally, the distance measure is not well specified, especially for minimum rent households. There is good evidence that, as one might expect, minimum rent households could rarely meet the requirement exactly. Indeed, they generally exceeded it by substantial amounts ($25-$30 per month). Furthermore, it appears that the rent requirements induced households to accept units with higher prices (rent/value ratios) than they would normally. As a result, the real change in housing obtained by these households may have been about half the change in expenditures.[14] Until these factors are explicitly modeled, it is difficult to assess the reasonableness of the estimated models or characterize the participation decision in terms of households' valuation of required housing.

APPLICATION OF THE ANALYSIS TO POLICY EVALUATION

It is clear that programs without housing requirements, such as percent of rent programs, will enjoy the same, relatively high, participation rates accorded to general income transfers. Housing requirements, on the other hand, clearly reduce participation. This reduction in program coverage is directly linked to the stringency of the requirements and focused on households whose housing fails requirements (Table 6.7). Households that would normally meet requirements on their own participate at only slightly lower rates than households in programs without requirements, but at roughly four times the rate for households that would not normally meet requirements. Because of this, demographic groups that are less likely to meet requirements are also less likely to participate in a housing gap program than in a similar unconstrained transfer program. Thus, for example, subsequent participation rates in the Demand Experiment were lower among minority households, large households, and households with very low incomes.

These patterns can be changed by changes in allowance payments.[15] Doubling the average payment level for the minimum standards requirement might have led to an overall participation rate of as much as 63% with a participation rate of about 50% among households that would not normally occupy minimum standard housing.[16] Alternatively, differences in participation by income group could have been eliminated by payment formulas involving much larger differences in payments to higher- and lower-income households. Specifically, the estimates in Kennedy and MacMillan (1980: chap. 4) suggest a contribution rate of 40%-45% of the first $4000 of annual income, with a rate of 25%-30% thereafter.

Such manipulations would not, however, change the fact that within each income class the program would still be less likely to serve the households that were least likely to meet the minimum standards requirements on their own. Very poor applicants are likely to include more households in poor housing than higher-income applicants. The connection, however, is not strong enough to overcome the tendency for households that are already in or about to move to required housing to participate much more readily than other households.

Program participation also has important implications for program ef-

TABLE 6.7 Effect of Housing Requirements on Participation

	Requirement Stringency			
	None	*Low*	*Higher*	
	Percent of Rent and Unconstrained	*Minimum Rent Low*	*Minimum Rent High*	*Minimum Standards*
Stringency				
Percentage of population that would normally meet requirements	100	68	40	31
Participation Rate				
All households[a]	84	61	44	37
For those that would normally meet requirements[b]	84	78	78	78
For those that would not normally meet requirements[c]	NA	23	20	19

SOURCE: Kennedy and MacMillan (1980: Tables 2.4, and 4.6).

a. The product of the acceptance and subsequent participation rates.

b. This is simply the acceptance rate, since all enrolled households that meet requirements subsequently participate.

c. This rate is defined as:

$$r = \left(\frac{P_E - P_N}{1 - P_N}\right) \bullet P_A$$

where

P_A = the acceptance rate;

P_E = the subsequent participation rate for all households that did not meet requirements at enrollment;

P_N = the estimated probability that these households would have met requirements normally, without the allowance program.

The rate, P_N, is estimated from equation 6.1 and Table 6.6, by

$$\ln\left(\frac{P_E}{1 - P_E}\right) = \ln\left(\frac{P_N}{1 - P_N}\right) + \hat{b}_E$$

where $\hat{b}_E$ is the experimental effect on participation in Table 6.6.

fects on housing. Most obviously, programs do not help households that do not participate. In addition, however, the effect of an allowance program on recipient housing would also be expected to depend on which households participate. A housing gap form of housing allowance is distinguished from a program of expanded unrestricted welfare payments only by its housing requirements. Households that would normally occupy housing that meets the requirements are effectively unconstrained; they are not required to select different housing from that which they would normally occupy or would occupy under a similar unconstrained program. But the housing changes generated by unconstrained programs were generally modest (Friedman and Weinberg, 1980b). Accordingly, most of the change in housing (apart from reduced rent burdens) generated by an allowance program, and all of the change beyond that generated by a similar unconstrained program, would be expected to come from the recipients that are constrained by the housing requirements to change their housing.

On the other hand, while the findings of the Demand Experiment with respect to relative participation rates under different programs are reasonably strong, there are several issues involved in the projection of findings to ongoing programs. First, the acceptance rates presented in Table 6.1 are clearly too high. Analysis of acceptance in the Demand Experiment was based on households that had been individually approached for enrollment and had received a brief (ten-minute) explanation of the program, including an estimate of their potential allowance payment. This was appropriate for analysis of program differences but would be expected to overstate acceptance rates in an ongoing program. Not only would no ongoing program conduct door-to-door outreach for all eligible households, but about 14% of the households contacted for enrollment refused to be interviewed while another 6% broke off the interview after hearing about the program (but before receiving a payment estimate). Thus, if we had used all contacted households as the base for acceptance, we would have calculated acceptance rates equal to about four-fifths of the rates in Table 6.1 (e.g., about a 67% participation rate for percent of rent households instead of 84%.[17]

The Demand Experiment was designed to compare alternative programs, not to estimate the absolute level of acceptance. The key finding of the analysis was that acceptance rates were quite similar across the different programs and that different programs did not have materially different enrolled populations. This fact allows us to interpret the analysis of subsequent participation in terms of the effect of housing requirements. Estimates of the absolute level of acceptance are better acquired through demonstrations. Fortunately, such demonstrations were carried out. The Supply Experiment provides direct observation of participation in an ongoing universal entitlement program. Likewise, the Administrative Agency Experiment yields information on the effects of outreach in limited entitlement programs.

The second problem in the interpretation of the Demand Experiment analysis arises from issues of population dynamics. At first glance, this problem is fairly straightforward: Low-income populations show considerable turnover; if meeting requirements takes time, then the participation rate in a program will be lower than the probability that households eventually participate to the extent that recently eligible households have not yet had time enough to participate. The first question, then, is the extent to which analysis of the cohort of eligibles enrolled in the Demand Experiment can replicate an ongoing program in which new households are constantly being added to the eligible population.

As it turns out, the cumulative participation rate for the cohort enrolled in the experiment will equal the current participation rate under certain fairly general conditions. If, for example, we have a steady state situation in which the total number of eligibles is fixed with a fixed distribution of eligibility duration, and if the probability that a household participates is a function of the time it is eligible, and once a household becomes a participant, it continues to participate as long as it is eligible, then the cumulative cohort participation rate will approach the current population participation rate of an ongoing program.[18]

Finally, the rules of the Demand Experiment provided that once a household met the housing requirements in a unit, it continued to meet requirements as long as it remained in that unit. Thus, a household could only go from meeting to not meeting requirements by moving. This rule was adopted for a variety of reasons, mostly reflecting the limited duration of the experiment. It seems quite likely, however, that an ongoing program would impose reinspection requirements for households that did not move. Furthermore, evidence from the Supply Experiment, where reinspection was required by the program, indicates that most households will respond by repairing defects (Rand Corporation, 1978: 69-73).

Should we therefore revise our impact estimates? The answer is probably not. As might be expected, the supply data show that the repairs involved to correct defects found on reinspection were generally trivial. In addition, analysis of initial upgrading in the Demand Experiment not only confirms that upgrading to meet minimum standards involved generally trivial costs, but suggests that it may not have increased total maintenance at all, but only focused it on items included in the standards, though the evidence is admittedly weak (Merrill and Joseph, 1980).

The major problem in the application of the analysis to policy evaluation is why so few households participated. It is here that the relatively loose specification used in the analysis breaks down. The findings of the Demand Experiment in the analysis of participation and of housing change make it clear that housing allowances are unlikely to have any substantial impact on the overall stock of housing. It is also clear that within feasible payment levels, allowance programs are unlikely to serve most eligible households in substandard housing. The central policy question that remains is simple: Does this mean that the allowance program is a failure? If the goal is to improve the housing stock or eliminate substandard housing, the answer is clearly yes. But this may not be the goal. Indeed, it seems likely that the goal is to offer all households, regardless of income, access to decent housing at reasonable cost. In this case, the program fails only if it did not offer such access. Furthermore, the nature of the failure can indicate important potentials for program redesign.

Consider, for example, the following reasons for nonparticipation. Suppose a household is simply not interested in housing and is unwilling to spend even 20% of its income on rent. If this household does not participate, it is at least arguable that the housing allowance did not fail but succeeded in targeting assistance to those most in need of housing assistance. Now change the story slightly and say that the household is not willing to spend even 20% of its income on rent because it has high medical bills due to some catastrophic illness. The argument that the household had a choice is weakened. But what if the cause is not illness but money set aside to put a child through college? What if it is money set aside for wine, women, and song? In part, the judgment of adequate access depends on the existence of other programs to provide access to other meritorious goods and services and/or on the way in which income is computed in calculating the allowance. Thus, for example, the allowance programs in the Demand Experiment did deduct major medical expenses from income; on the other hand, they included educational grants, but not food stamps.

Now consider another case. We know that housing prices vary. Households may end up paying quite different amounts for similarly located units with similar amenities. Say that a household pays an unusually low rent, so much so that it is not worth its while to accept the allowance and find another unit that meets program requirements. It is hard to argue that the allowance has failed such a household when it does not participate. But now take the other side of the coin. Say the household cannot find a unit that meets requirements at a rent low enough to be affordable, even with the allowance. Then the program would appear to have failed. The analytic specification used in the analysis of participation is simply not powerful enough to sort out nonparticipation due to tastes, other exigencies, good deals, or problems in search. We do know that the overall participation rates, particularly under minimum rent, seem to make sense only if one takes account of the actual rents that households ended up paying to meet minimum rent requirements and that meeting these rent requirements induced households to accept higher-priced units (that is to increase rent more than housing services).[19] But we cannot now sort out the different factors involved.

These issues of interpretation require the specification of much more powerful models of individual search and more detailed characterization of individual circumstances. While experimental designs can employ relatively loose models to estimate overall program impacts, the interpretation of the results necessary for final policy judgments is still strongly model-dependent.

NOTES

1. The exact static expression for the minimum payment needed to compensate a household for obtaining a given level of housing consumption is:

$$C_R = \begin{cases} R_R - R_N - \int_{H_N}^{H_R} p(H, U_N)dH & \text{if } H_R > H_N \\ 0 & \text{if } H_R \leq H_N \end{cases}$$

where

C_R = the minimum payment needed to compensate a household for obtaining at least level H_R in housing;

H_R = the required level of housing consumption;

H_N = the household's normal housing consumption;

R_R, R_N = the rents associated with required and normal housing consumption, respectively;

$p(H,U_N)$ = the inverse compensated demand function, evaluated at the household's normal (nonparticipating) utility level, U_N.

Because this expression involves movement along an indifference curve, a variety of theorems on the signs of various derivatives with respect to the level of requirements, housing prices, and income follow. (For details, see Kennedy et al., 1977, chap. 3, or Kennedy and MacMillan, 1980, Appendix XVII.) If general costs, C_p, are nonmonetary, the value of C_P should be added to income in deriving H_N and U_N. In addition, as is usual with composite commodities, theorems for a specific list of physical requirements are limited to characterizations of proportional shifts in requirement levels.

2. The problem is that the condition of participation is that the payment (S) exceed general costs of participation, C_P (such as the bother of income reporting and housing inspections) and that this quantity in turn be greater than the expression for C_R in the previous footnote. Note, however, that normal rents greater than required levels ($H_R \leq H_N$) will not, of course, compensate general participation costs. As a result, the probability of participation involves the product of two probabilities (e.g., that $S - C_P$ be positive and $H_R \leq H_N$, or that $S - C_P - C_R$ be positive and $H_R > H_N$). This does not, of course, preclude estimation of a single probability function, but it does make tracing the theoretical connection between the individual terms and the estimated coefficients difficult.

3. See Kennedy and Merrill (1979) for evidence of price heterogeneity and shopping effects.

4. Terms in the estimated residual from a regression of enrollment rents on various demographic descriptors, including interactions with moving probabilities, were insignificant and

were dropped from the final specification. This is not theoretically impossible, since the relevant normal rent may be an average of rents over future periods.

5. The formula for the expected cost of moving to meet requirements (C_m) is

$$C_m = -\ln (p_m)$$

where p_m is the estimated (logistic) probability that the household would move normally. This is derived from a model in which a household moves if its net cost of moving (V) is negative, where V is a function of observed characteristics (x) and a stochastic term c. Thus,

$$C_m = (X + E(c|X + c > 0))(1 - p_m).$$

If c has a logistic distribution,

$$p_m = [1 + e^x]^{-1}$$

and the formula for C_m follows.

6. Differences in the incidence of concerns in Phoenix may reflect the fact that the allowance program income limits were higher in Phoenix than in Pittsburgh, while income limits for other housing programs were generally lower. Thus, households in Phoenix more often might have been surprised at the idea that they were eligible for a government transfer.

7. The chi-square statistics for pooling housing gap, percent of rent, and unconstrained treatments (up to a constant) were 19.73 with 14 degrees of freedom and 13.73 with 15 degrees of freedom in Pittsburgh and Phoenix, respectively.

8. The term in $\ln (p_M)$ in theory arises only if the household cannot more easily arrange to meet requirements in its original unit instead of moving. Merrill and Joseph (1980) found that, as might be expected, such in-place participation was generally confined to households that were relatively close to meeting requirements. Thus, as an alternative specification, $\ln (p_M)$ was entered only for households with values of $R_R - R_O$ greater than \$15 (based on Merrill and Joseph, 1980: Table 3.6). This specification again generally showed no significant effects for $\ln (p_M)$.

9. The chi-square statistics for dropping demographic variables from an equation stratified by type of housing requirements were 15.11 with 10 degrees of freedom (Pittsburgh) and 16.82 with 11 degrees of freedom (Phoenix).

10. The only case in which there was a significant difference was for minimum rent low households in Phoenix, for which the (chi-square was 8.46 with 2 degrees of freedom). This reflects a significant and negative distance effect for control households as compared with an insignificant distance effect for housing gap minimum rent low households.

11. The chi-square statistics for pooling the sites were 6.57 (minimum standards), 2.74 (minimum rent low), and 2.12 (minimum rent high), with 3 degrees of freedom in each case.

12. Other coefficients were not materially different from those in Table 6.5 (see Kennedy and MacMillan, 1980: Table 4.5).

13. This idea was originally suggested to us by Henry Aaron.

14. See Kennedy and MacMillan (1980: Appendix VIII).

15. Payment amounts only had statistically significant effects for minimum standards households. The payment effects for minimum rent programs, however, while not significant, were also not significantly different from those estimated for minimum standards. See Kennedy and MacMillan (1980: Table 4.4).

16. This estimate applies to a situation where the eligible population is held constant while payment levels are doubled.

17. On the other hand, households approached for enrollment in the Demand Experiment had already been interviewed twice—first in a 15-minute screening interview and then in an

hour-long baseline interview. Some households that refused to be interviewed may have been refusing another interview and not turning down an allowance program. Likewise, the experimental programs were new, without the established (good or bad) reputation of an ongoing program.

18. There are, however, several caveats to this happy coincidence. See Kennedy and MacMillan (1980: Appendix VII).

19. See Kennedy and MacMillan (1980: chap. 4).

7

The Role of Supportive Services

WILLIAM L. HOLSHOUSER, Jr.

□ IN ITS SIMPLEST FORM, a housing allowance is a cash transfer intended to improve the housing conditions or reduce the rent burden of its recipients. This simplest form of the program, however, is rarely proposed. Negative constraints are added, such as a requirement that the subsidy amount be spent on housing, or a prohibition against subsidizing "substandard" units. Positive additions are also frequently proposed, among them a requirement that administering agencies provide "supportive services"—for example, program information, counseling, legal services, transportation, child care, assistance in locating available units or negotiating with landlords—to enable participants to make better use of their allowances.

In the planning stages of the Administrative Agency Experiment, some researchers advocated testing a cash transfer program in which recipients were required to purchase housing of decent quality but were not offered agency assistance to do so. If this approach proved feasible, of course, it promised to be less costly than a program providing both allowances and services.

Other researchers asserted that supportive services of some kind might be necessary for the program to work at all, or for it to work equitably. This disagreement reflected a lack of clarity in social science literature about the value of counseling and other supportive services, not only in housing but also in social programs generally. Hartmann and Keating (1974) reflected the convictions of many service advocates when they wrote, "It is inconceivable that low-income tenants armed only with housing allowances can succeed in the housing market, given the current state of landlord-tenant legal relationships throughout most of the country." They advocated legal reform in the long run and counseling and other services in the interim. Skeptics,

---

AUTHOR'S NOTE: *Portions of this material have been published previously in Holshouser (1977) and Hamilton (1979).*

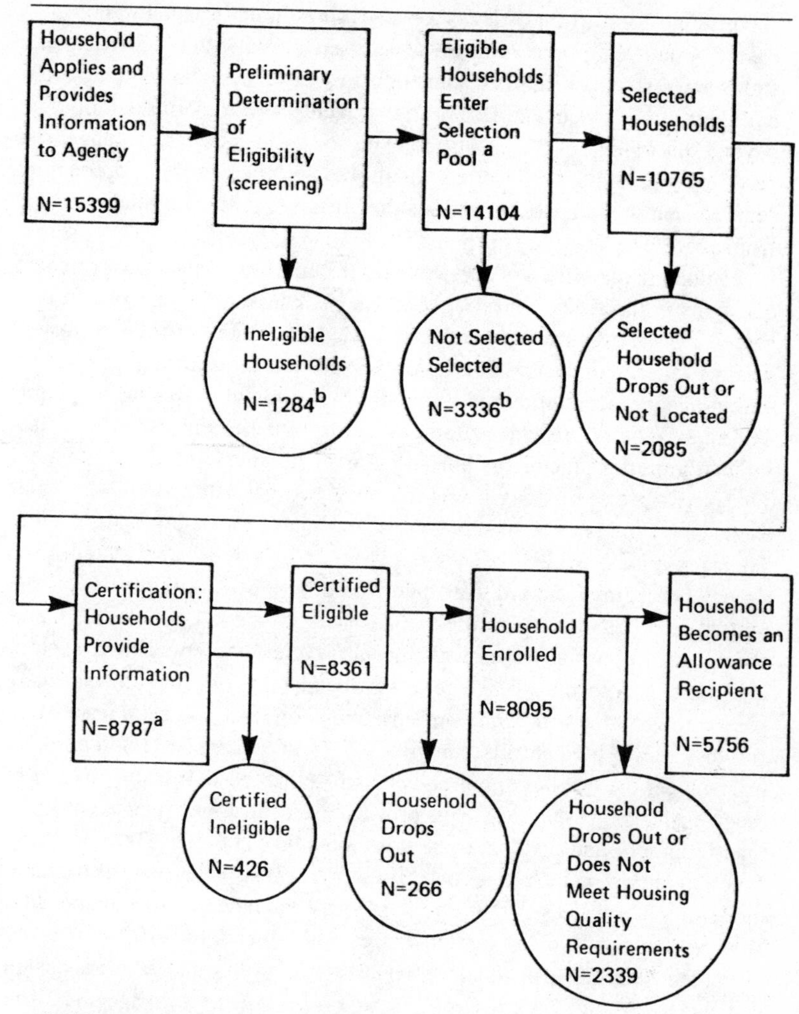

SOURCE: Hamilton et al. (1977b: 10).

a. In some cases (mostly in Bismarck), certification took place before selection: A total of 107 households were certified eligible at various sites but were never selected.

b. Missing cases: eligibility of applicants, 11; selection, 3.

FIGURE 7.1 Steps in Participation in the AAE

meanwhile, were able to point to a general absence of positive findings in previous attempts to evaluate counseling or other services (see, for example, U.S. General Accounting Office, 1973b). No one doubted that some services could help some people, but no clear case could be made for the effectiveness of services in general or of specific services that the agencies might offer.

In the end, agencies offering housing allowances were required to provide "counseling" and to make legal assistance available in equal opportunity cases. "Counseling" was to cover at least an introduction to participants' rights and responsibilities and some information about local neighborhoods and their housing markets. However, the agencies were permitted to go far beyond this minimum if they wished to do so. As a result, the eight agencies participating in the AAE varied widely both in their overall approach to services and in the specific services they offered. Agency attitudes ranged from aggressive advocacy to "hands off."

Evaluating the effects of services thus became one of the major research questions in the AAE. And despite some weaknesses, the experiment presented a good opportunity to address the question. The primary weakness was the reliance on "natural variation." There was no control group and no within-agency variation based on random assignment to treatment groups. The analysis therefore had to attempt to control for differences between sites in such important factors as housing market tightness and discrimination. The AAE's greatest strength was the ability to place trained observers at each agency for a period of about a year and a half of start-up and operation. Their presence provided the analysis with a much richer understanding of agency procedures and attitudes than would have been possible with intermittent site visits, and permitted researchers to suggest hypotheses about potentially important issues with the observers, who could then explore them on-site. A second advantage was the availability of a relevant and easily measured outcome—the program's participation rate. The main purpose of services was to help enrollees qualify for program benefits. Each enrollee had to meet the housing requirements before he or she became eligible to receive a housing allowance. This proved easier for some types of enrollees than for others, and easier at some AAE sites than at others. By controlling for other important influences on enrollee success rate, the analysis was able to identify a "services effect" and to identify the types of participants who benefited from it, and the circumstances under which the effect had taken place. We were thus able to address questions of policy and program design and also to produce findings useful to administrators of local agencies. The analysis is also of interest on methodological grounds, especially in its complementary use of qualitative and quantitative data.

BARRIERS TO PARTICIPATION IN THE AAE

Figure 7.1 shows the steps involved in becoming a recipient in the AAE. The services analysis dealt largely with the success or failure of enrollees in becoming recipients.[1] There were 8095 enrollees, of whom 5756 (71%) became recipients. Variations in this "success rate" among the eight AAE agencies and across participant characteristics provided the primary indicator of service effectiveness.[2]

Enrollees were given 60 days, with a 30-day extension at agency option, to qualify as recipients. This required that three things occur. First, the enrollees had to locate a rental unit they wanted to live in. Second, the landlord had to agree to sign a lease containing several provisions required by the program. Third, the unit had to pass agency inspection. The primary goal of the supportive services was to help enrollees meet these requirements. The extent to which they were able to do so varied with four other factors influencing enrollee success rate: enrollees' attempts to move or stay in place, housing market tightness, the quality of local rental stock relative to the agency's housing quality standards, and the amount of discrimination encountered by enrolled households. These other factors will be discussed first.

Participants were permitted to stay in the units they were living in at enrollment if the landlord would sign the lease[3] and if the unit could pass inspection.[4] Households were asked at enrollment whether they planned to move or stay. About 84% of those planning to stay became recipients, and the proportion varied little from agency to agency. Households planning to move fared less well. Overall, only 62% became recipients. Further, the success rate for this group differed considerably across locations, ranging from a high of 83% in Tulsa to lows of 52% in Peoria and 28% in Jacksonville.[5]

It had been widely speculated that housing allowances will not work well, at least in the short run, in tight housing markets.[6] Peabody (1974), one of the early advocates of the housing allowance experiments within HUD, wrote that housing allowances "cannot be used in tight markets where vacancy rates are less than five to six percent without some parallel program to expand production. Otherwise there will be an imbalance of supply and demand and rents will be driven up without improving overall housing quality." Other writers expressed the opinion that even a small-scale program could not work well where decent units were in short supply (Hartmann and Keating, 1974; Weaver, 1975; U.S. Comptroller General's Office, 1974).

The AAE included four cities that had relatively tight housing markets—those with vacancy rates estimated to be 4% and 6% during the time most enrollees were searching for housing—and four with looser markets—with vacancy rates of about 8%–13%. Analysis showed market tightness to be an important determinant of enrollees' success in becoming recipients. About 60% of the households operating in the tighter markets became recipients, compared with about 85% in looser markets. Not surprisingly, market tightness was most important for enrollees who attempted to move.

AVAILABLE HOUSING COMPARED WITH AGENCY STANDARDS

Agencies were required to inspect units to assure that recipients occupied "decent, safe, and sanitary" dwellings. Quality standards were locally determined. At some sites, inspections were done by the participants' counselors, while at others they were done by the city's code enforcement staff or other

TABLE 7.1 Enrollee Success Rates by Stringency of the Agency Housing Standard and Market Tightness (percentages)

Market	Agency Requirements: Less Stringent	Agency Requirements: More Stringent
Looser	86	84
Tighter	65	57

SOURCE: Holshouser (1977: 14).
Sample size=8095.

TABLE 7.2 Enrollee Success Rates for Attempted Stayer Households, by Unit Quality and Stringency of Agency Requirements (percentages)

Unit Quality[a] *(Adjusted Rent)*	*Agency Requirements*: *Less Stringent*	*Agency Requirements*: *More Stringent*
Lower	81	74
Higher	87	83

SOURCE: Holshouser (1977: 14).
a. Lower unit quality is indicated by actual gross rents at enrollment of less than 70% of the amount estimated to be the typical cost of housing that would meet the agency's standard (for a household of that size in that market). Higher unit quality is indicated by a gross rent of 70% or more of the estimate.
Sample size=3117.

professional inspectors. There was considerable variation, too, in the quality of the housing stock itself from one city to another. Some agencies defined and enforced quality requirements which excluded a large proportion of the low-cost housing stock, while others applied standards that would admit most local units. (Either course could be perfectly appropriate, of course, depending on the general quality of the stock.) In these circumstances, enrollees in the more stringent communities might be expected to have a lower rate of success in becoming recipients. The lower success rate might result because fewer enrollees would already be living in acceptable housing, and the many attempting to move would find it more difficult to find acceptable housing.

This prediction proved accurate in the AAE. Enrollees at sites with more stringent standards[7] succeeded less frequently than those at other sites. Table 7.1 compares the percentages of enrollees that became recipients under different degrees of agency stringency and market tightness. Success rates varied from a low of 57% in tighter market, higher-stringency sites to a high of 86% in sites with lower stringency and looser markets.

The quality of enrollees' original housing interacted with the effect of agency quality requirements.[8] Where the quality of the original housing was poor and the agency standard was stringent, the failure rate for enrollees who wanted to stay in their units was high. Table 7.2 shows the outcomes for enrollees who wanted to stay in their preprogram units. Among those in better housing at enrollment at agencies with less stringent requirements,

87% became recipients; the rate for those in relatively poorer housing facing more stringent requirements was 74%.

Fears have often been expressed that housing allowance participants subject to discrimination might find it more difficult to rent acceptable housing in desirable neighborhoods. For example, historic patterns of segregation might mean that disproportionate numbers of black enrollees would be living in substandard housing at enrollment. But current discrimination might make it difficult for blacks to rent better housing, even with the additional subsidy money. If they could not rent adequate housing, they would be forced to drop out of the program.[9]

The only ethnic groups represented in significant numbers in the AAE were whites (non-Hispanic), blacks, and Hispanics. Of these groups, only blacks had more difficulties than other groups, especially in sites with tight markets. Overall, 53% of the black enrollees became recipients, compared to 77% of the whites. And as Table 7.3 shows, black enrollees planned to move more often than whites. Blacks tended to start out in lower-quality housing (presumably because of both lower incomes and market discrimination), and therefore had to move more frequently to qualify for payments. But even when moving plans and other household demographic characteristics are taken into account, the black success rate is lower and impossible to explain without reference to racial discrimination.[10]

Several other factors related to enrollees' chances of becoming recipients deserve mention, but they were less influential than those already discussed. Large families became recipients less frequently than others, perhaps because of a scarcity of large units or market discrimination against such families. In general, the elderly tended to be more successful than the nonelderly, largely because they were more often able to stay in their original units. None of these factors caused large enough differences in enrollee success rates that a program administrator would be likely to seek corrective action.

THE EFFECTIVENESS OF SERVICES IN ALLEVIATING BARRIERS TO PARTICIPATION

The AAE agencies offered many types of services. Before enrollment, all agencies provided selected applicants with information about program rights and responsibilities in order to permit participants to make informed decisions about whether to enroll. After enrollment, a housing and equal opportunity counseling session was offered. Attendance at postenrollment sessions was voluntary in some locations and mandatory in others. Some agencies presented group sessions, while others scheduled individual meetings with counselors. Four of the agencies (Salem, Springfield, San Bernardino, and Bismarck) adopted overall approaches which were characterized (on the basis of on-site observer descriptions) as "enabling." These agencies

TABLE 7.3 Moving Plans and Success Rates for Black and White Enrollees

	Black	*White*
Tight Market Sites		
Percentage planning to move	73%	49%
Percentage becoming recipients	47	68
Loose Market Sites		
Percentage planning to move	73	41
Percentage becoming recipients	82	86

SOURCE: Holshouser (1977: 17).
Sample size=8095.

viewed the enrollee as an independent agent, or as capable of becoming one with training. This led them to focus on preparatory services, delivered to enrollees before housing search, which anticipated problems the agency expected participants to encounter. Two agencies developed "casework" approaches, with counselors often assuming much of the responsibility for locating available units and negotiating with landlords. In these agencies, services were concentrated on the housing search process and sought to alleviate the specific problems encountered by individual enrollees, rather than anticipating them. The final two agencies were characterized as laissez-faire in their approach to services. Little emphasis was put on services, either preparatory or problem-oriented. Participants were told that services were available but were left to request them if they needed or wanted them.

A major decision for researchers was determining which aspects of services and service delivery to analyze. There was little theory to indicate which of many differences should be considered important: groups versus individual sessions? mandatory versus voluntary attendance? preparatory versus problem-oriented services? emphasis on legal assistance and equal opportunity versus a "low profile" on protection of enrollees' civil rights? the availability of such specific services as transportation or child care to attend counseling sessions or to search for housing versus leaving the provision of these matters to the enrollees themselves? It became evident that taxonomies were needed which captured aspects of the agencies' overall approaches but which also reflected the types of services offered. After considerable preliminary analysis and several meetings with on-site observers, it was decided to distinguish between "formal" services and "responsive" services and to classify an agency's emphasis on each type of services as high or low. On-site observers agreed that these categories captured most of the variation in details of attitude and procedures that they considered intuitively important on the basis of their site experience.

Formal services were generally standardized and presented to enrollees shortly after their enrollment. Agencies provided responsive services to individual enrollees as particular problems arose.[11] Although all agencies of-

TABLE 7.4 Market Tightness, Agency Type, and Emphasis of Formal and Responsive Services

Site	*Housing Market Tightness*	*Agency Type*	*Emphasis on Formal Services*	*Emphasis on Responsive Services*
Salem	Loose	Enabling	High	Low
Springfield	Tight	Enabling	High	High
Peoria	Tight	Laissez-faire	Low	Low
San Bernardino	Loose	Enabling	High	Low
Bismarck	Loose	Enabling	High	Low
Jacksonville	Tight	Laissez-faire	a	Low
Durham	Tight	Casework	Low	High
Tulsa	Loose	Casework	High	High

SOURCE: Holshouser (1977: 20).

a. Jacksonville planned and offered an extensive program of formal workshops. Unlike any other sites, however, this formal content was offered in voluntary, not mandatory, sessions. Only about a quarter of the agency's enrollees attended. Jacksonville's emphasis on formal services was therefore high in intent but low in fact.

fered both formal and responsive services, emphasis differed. Some agencies concentrated on formal information and training and seldom intervened in enrollees' search efforts; some gave less "front-end" information but provided extensive resources or direct help to households encountering problems. Still others offered a high level of both types of services. Table 7.4 shown the agencies' relative emphasis on formal and responsive services.

FORMAL SERVICES

Variations in the level of formal services do not seem to have affected enrollees' chances of becoming recipients. In loose housing markets, even the lowest level of services was adequate. Success rates in these markets were consistently high, no matter what services the agencies offered. In loose markets, then, a low level of formal services appears to cause no loss in effectiveness, and, as will be discussed, it offers an opportunity to hold down administrative costs.

It is more difficult to assess the value of formal services in tighter markets. Because no tight market agency combined a high level of formal services with a low level of responsive services, the available data do not allow a determination of the importance of formal services alone under such conditions. Fortunately, Springfield offered a partial exception to this problem. The Springfield agency kept careful records of enrollees' attendance at formal sessions and their use of responsive services. Springfield enrollees who attended only the mandatory formal sessions succeeded at a significantly lower rate than those who also attended voluntary training sessions or used individual responsive services. Among households that planned to move, 52% of those that received only formal services succeeded in becoming re-

TABLE 7.5 Success Rates for Enrollees Becoming Recipients Under Various Service and Housing Market Conditions, by Race

		Moving Plans				
		Move		*Stay*		
Market Conditions	*Level of Services*	*Black*	*White*	*Black*	*White*	*Totals*
Tighter	High (Springfield, Durham)	290 65%	336 63%	223 77%	470 77%	1319 70%
	Low (Peoria, Jacksonville)	219 26%	389 54%	92 53%	552 76%	1252 51%
Looser	High (Tulsa)	132 82%	211 84%	38 84%	479 88%	860 86%
	Low (Salem, San Bernardino, Bismarck)	67 78%	779 81%	40 85%	1063 89%	1949 85%
Total		708 47%	1715 69%	393 71%	2564 84%	5380 71%

SOURCE: Holshouser (1977).

Sample size=(enrollees) 7719. Most missing cases are enrollees in other ethnic groups or those who were undecided about their plans to move.

NOTE: This table does not include stringency of housing quality requirements because of data limitations for that variable.

cipients, compared to an 80% success rate among those that received additional services. These findings are not conclusive, but they suggest that formal services alone, even if provided at a high level, are insufficient to ensure a high enrollee success rate in tighter market conditions.

The Durham agency also operated in a tighter housing market. It provided minimal formal services to enrollees but offered extensive responsive services. The proportion of enrollees becoming recipients in Durham was essentially the same as in Springfield, where both formal and responsive services were offered at high levels.

These examples suggest the feasibility of holding formal services to low levels, even in tight markets. The AAE data are too limited to consider the findings conclusive, but the potential savings of such a policy recommend it for further research.

RESPONSIVE SERVICES

Because formal services had little visible effect in the experiment, analysis of service effectiveness came to focus largely on responsive services. In tighter housing markets, higher levels of responsive services were associated with higher levels of enrollee success. Table 7.5 shows the proportion of enrollees who became recipients under conditions defined by moving intentions, market tightness, race, and service level. The percentage of enrollees becoming recipients varied from a low of 26 for blacks who wanted to move in tighter market sites with few services to a high of almost 90 for whites in looser markets who wanted to stay.

Only one of the looser market sites, Tulsa, offered a high level of responsive services. The others (Salem, San Bernardino, and Bismarck) offered largely formal services. Because the success rates of the four agencies are so similar, it seems likely that the high level of services in Tulsa did not add significantly to enrollees' ability to become recipients.

The results observed in the tighter market sites are more striking. Two agencies (Springfield and Durham) offered a high level of responsive services, and the other two (Peoria and Jacksonville) offered much lower levels. These variations in service level are clearly associated with variations in enrollee success rate: 70% of the enrollees in tight-market sites with high levels of responsive services became recipients, but only 51% succeeded where service levels were lower.[12]

Services made a difference for households planning to move in tight housing markets. Their effect was significant for whites but crucial for blacks. High levels of responsive services improved the success rate for whites by 9 percentage points; for blacks the improvement was 39 percentage points.

The level of services made less difference for households that wished to stay in their original units, even in tight markets. However, 77% of black households in this category became recipients at higher service sites, compared with 53% where services were lower.

With market tightness and moving plans held constant, the high services appeared to help female-headed households more than male-headed households and those with welfare income more than those without it. The patterns were weak, however, and not as solidly based as the effects found for black enrollees. Service levels did not help the elderly more or less than the nonelderly, or large households more or less than smaller ones.

These findings were consistent with observational data and with participants' responses to survey questions. Both show that the high-service, tighter market agencies responded to the special problems of households at a disadvantage in the housing market. Springfield, for example, emphasized equal opportunity training and aggressive use of legal services. Durham interceded with landlords, helped enrollees negotiate leases, and followed problem cases closely. Both programs apparently worked.

Responsive services to enrollees therefore seem compensatory. Enrollees in tight housing markets, enrollees planning to move, and black enrollees had difficulty becoming recipients; it was precisely these groups responsive services helped most. Services did not, of course, completely equalize success rates. Enrollees in loose markets did somewhat better than enrollees in tighter markets, even with high responsive services. In tight markets, enrollees planning to move were still less likely to become recipients than those planning to stay in their preprogram units. But services reduced those differences, and the dramatic disparity between outcomes for black and white enrollees in low-service sites was almost nonexistent in the high-service sites.

COST IMPLICATIONS

The findings suggest that high levels of formal services may not be necessary in looser market sites, and perhaps not even in tighter market sites. The evidence also indicates that low levels of responsive services are adequate for looser market sites but that more extensive services have definite advantages in tighter market sites.

These conclusions might have important consequences for the administrative costs of a housing allowance program. Services to enrollees accounted for about 37% of all the direct administrative costs of bringing families into the AAE program. Variation in these costs can therefore have an important effect on total administrative costs. Table 7.6 shows that services costs varied substantially by intensity and method (see Silberman, 1977, for a detailed analysis of AAE service costs). The table shows the estimated cost per enrollee[13] of varying the intensity of formal and responsive services and varying the way formal services are provided.

Within any given format of delivering formal services, the difference between lowest and highest levels of service was about $32 per enrollee in 1973 dollars. If, for instance, formal services were delivered in group sessions, costs ranged from $16 for sites with a low emphasis on both formal and responsive services to $48 for a high emphasis on both. The cost difference between a low and high emphasis on formal services was about $19, and the difference between low and high responsive services was estimated at $13.

The relatively large incremental cost of a high emphasis on formal services is particularly important given the indications that such services may not be necessary. Further research, perhaps in other areas of service delivery, would be desirable to determine more conclusively whether such services affect the achievement of program objectives.

TABLE 7.6 Estimated Labor Cost per Participant, by Level and Method of Services Offered, 1973

		Method of Formal Service Delivery		
Emphasis on Formal Services	*Emphasis on Responsive Services*	*Group*	*Individual*	*Group Plus Individual*
Low	Low	$16.41	$31.86	$45.15
	High	29.33	44.78	58.08
High	Low	35.12	50.57	63.87
	High	48.04	63.49	76.79

SOURCE: Holshouser (1977: 25).
NOTE: Figures include indirect costs at the median AAE rate.

The $13 increment in administrative costs for a high level of responsive services may be considered acceptable in tight markets simply because it will improve some groups' chances of participating in the program. In addition, the expenditures may reduce the money and administrative effort devoted to dealing with applicants who never become recipients.

Responsive services also appear to be one of the major administrative devices to reduce attrition costs. These costs can be substantial. AAE attrition costs amounted to 36% of all expenditures for bringing families into the program (Maloy et al., 1977). Responsive services can reduce attrition costs by helping more enrolled families qualify for allowance payments. In fact, under some circumstances, reduced attrition costs may offset the higher cost of responsive services, yielding a lower average administrative cost per family.

In summary, formal services were found to have little effect on success rates. In a nonexperimental housing allowance program, consideration might be given to presenting program information through audiovisual means, or perhaps on a regional basis, as a means of controlling unnecessary costs. Responsive services, however, had a tangible effect on program outcomes, enhancing some households' opportunity to participate. Moreover, the effect is compensatory. That is, responsive services provided the most help to the groups with the largest problems in the programs: black households and households attempting to find units in tight housing markets. In such markets, the effect is so marked that savings in attrition costs tended to offset the additional cost of providing the services.

NOTES

1. Issues other than becoming a recipient were studied. These included the effect of services on recipients' housing quality, on patterns of mobility and ethnic deconcentration, on maintenance of units under lease, and on terminations by recipients. The analysis of success in becoming a recipient, however, was the most central issue and was also the source of the most helpful findings. It is therefore the focus of this chapter.

2. All eight agencies tried to have all or most enrollees qualify for benefits. Some agencies administering the similar Section 8 Existing Housing program have adopted a deliberate strategy of overenrollment: The agency enrolls many more households than its number of available slots, then gives benefits to those who qualify soonest and terminates those who have not yet qualified at the time all slots are filled. No AAE agency adopted this approach, partly because of pressure from HUD not to overenroll, partly because they had trouble enrolling as many households as their plans called for, and partly because the agencies received administrative fees on the basis of its number of recipients, regardless of how many households it had enrolled, certified, or counseled.

3. The lease met with landlord resistance on at least two grounds. Many landlords had never used leases and did not wish to be forced to do so. Some landlords objected to a clause promising not to discriminate on the basis of race or national origin. Since there was no requirement to leave the unit in the program if the recipient household moved out, this clause did nothing to restrict a landlord's use of his or her unit. Nevertheless, many landlords disliked it.

4. If the unit failed an initial inspection, it could be repaired to comply with the quality

standards. However, since the program was a short-term demonstration, there was little incentive for a landlord to make extensive repairs or renovations. Some did so, but their motivation was not primarily economic.

5. Enrollees' plans to move or stay were reasonably accurate predictions of their actual behavior: Among those who became recipients, 87% followed the plans they had reported at enrollment (Wild, 1977a).

6. The "short run" here might be about six to nine years. On the supply side, Muth (1969) estimated that an urban housing market requires six years to make a 90% adjustment after being put out of equilibrium. Mayo (1977) estimated that once a housing market is put in disequilibrium, 75% adjustment to a new equilibrium would be made after five to six years, and 90% after eight to ten years.

7. Stringency depended on both the stated standards and how they were applied. The measure was derived by comparing deficiencies found in comparable items on agency inspection forms and forms completed by independent inspectors employed and trained by Abt Associates.

8. "Housing quality" was defined as the ratio of actual rent to the local cost (estimated by a local panel of experts using the Delphi method) of renting a unit of appropriate size in "modest standard" condition. This measure is highly consistent with other, more direct measures of unit quality.

9. On predicted problems see Shaw (1974), Dolbeare (1974), and Senate Committee on Banking, Housing and Urban Affairs (1973). On actual problems encountered in the Jacksonville AAE agency see Holshouser (1976) and Wolfe and Hamilton (1977).

10. These findings reflect to some extent extraordinary problems encountered by black enrollees in Jacksonville (Holshouser, 1976). But the patterns persist, though in less extreme form, if Jacksonville is dropped from the analysis.

11. See Trend (1977) for a more detailed discussion of the agencies and the services they offered.

12. The figures are influenced by a very low success rate for enrollees in Jacksonville; however, multivariate analyses excluding Jacksonville yield the same general result (see Temple and Warland, 1977; Wild, 1977b).

13. The costs shown are based only on labor costs, which were 82% of AAE costs. The percentage would have been higher if the cost of equipment and materials had been treated as a capital cost instead of a current expenditure. The latter treatment was necessary because of the short duration of the AAE. Thus, these costs underestimate the full cost of services. Procedures for estimating the direct costs are presented in Silberman (1977). Indirect costs have been estimated at 122% of direct costs (the median figure for the enrollment period in the AAE), and they are included in the figures shown in Table 7.6.

Part III

Housing Consumption

☐ CHAPTER 8, by Joseph Friedman and Daniel Weinberg, covers the ways housing allowances affected housing consumption. Their paper discusses the effects of the two different basic payment formulas that were tested in the Demand Experiment. One was the housing gap formula, where payments were based on the difference between the cost of modest standard housing and a fraction of the household's income. This formula was considered a likely formula for use in a national program and was also used exclusively in the other two experiments. Under most variants, payments using the housing gap formula were made only to households that met a housing standard.

The other payment formula, called percent of rent (a proportionate rent rebate), was included mainly to allow estimation of important parameters of the demand for housing—price and income elasticities. These were found to be low; a 10% price decrease was found to lead to a demand increase of only about 2%, while a 10% income increase led to a demand increase of about 4%. Given these modest findings and the finding that most participant households either lived in acceptable or almost acceptable housing at enrollment, one might therefore not be surprised at the actual finding of fairly modest allowance-induced housing changes. The chapter concludes that these increases were not significantly different from the increases that would have been caused by an unconstrained income transfer. Thus, it appears that housing requirements act as a barrier that limits participation and should be imposed only if the policymaker has a strong conviction about its specific importance in ensuring that low-income households live in adequate housing.

Chapter 9, by Sally Merrill, describes the use of hedonic indices in EHAP. Hedonic indices attempt to sort out the influence of housing and nonhousing factors in determining the market value or rent of a housing unit. The indices can be constructed to be sensitive to both unit and neighborhood characteristics and at the same time to remove the effects of nonhousing factors such as inflation, tenure condition, or racial discrimination. In the experiments, hedonic indices were thus used to analyze experimental effects on housing "quantity" as opposed to expenditures, and for a variety of housing market analyses.

Merrill's chapter first describes how hedonic indices were used in HADE to address the question of whether allowance-induced changes in rental expenditures led to concomitant changes in real housing services and the way the allowances altered tenant behavior in shopping for housing. It then describes the use of hedonic indices to estimate the demand for housing components, such as space and quality, using information from the Supply Experiment. Next, a more detailed breakdown of housing into numerous attributes is described and applied to an analysis of racial segregation in the two HADE sites.

Chapter 10, by John Mulford, summarizes findings on housing consumption changes and changes in the budget allocations of housing allowance recipients in the Supply Experiment. Unlike HADE, HASE included both renters and homeowners. Renters improved their housing partly by moving and partly by repairing their initial units; owners achieved their housing improvement almost exclusively by repairing their homes. The chapter estimates the program's effect on recipients' housing consumption and housing quality and discusses how those effects occur. It then examines how recipients allocate their allowances between housing and other consumption. Finally, the chapter compares housing allowances with public housing and unrestricted cash grants in terms of consumption increase per program dollar spent.

Chapter 11, by Peter Rydell and Lance Barnett, presents the findings from the Supply Experiment on the way a full-scale housing allowance program would affect the price of housing. Before the experiment, many economists feared that a housing allowance program, which is designed to stimulate housing demand without affecting housing supply directly, would, at least in the short run, cause inflation in the price of housing. Predictions of increases in the range of 10% or more were common. The price increases found in the Supply Experiment were much lower. In fact, in both housing markets, price increases were less than increases in the costs of operating rental property. Within each market, rents for recipients' dwellings increased more modestly than rent in nonrecipients' dwellings.

The authors present an explanatory model of how an allowance program affects demand, how supply responds to increased demand, and how the consequent price change is determined. The reasons for the low price change are traced to several factors that were not realized at program initiation in the early 1970s: (1) Participation in the program was quite low and spread out over time; (2) a large fraction of the participants lived in acceptable housing even before they joined the program; (3) some upgrading of substandard units occurred in place and was relatively inexpensive; and (4) in the short run, occupancy rates increased to absorb large parts of the demand shift.

8

Rent Rebates and Housing Standards

JOSEPH FRIEDMAN and DANIEL H. WEINBERG

☐ THE DEMAND EXPERIMENT tested two different payment formulas—housing gap and percent of rent. The housing gap formula, wherein payments were based on income and an estimated cost of standard housing, was considered the most likely formula for use in any national housing voucher program. In contrast, the percent of rent formula, a proportional rent rebate, was included primarily for analytical reasons—from the way households responded it was possible to estimate the sensitivity of housing demand to variations in housing prices.

This chapter has two purposes. First, it presents evidence on price and income elasticities of housing demand based on a controlled social experiment—the Housing Allowance Demand Experiment. Next, it estimates the experimental effect on housing consumption of a constrained income transfer—one conditional on the assisted household occupying adequate housing. Finally, we make some concluding remarks. Interested readers are referred to Friedman and Weinberg (1980a-1980c, 1981, 1982) for more details.

THE PERCENT OF RENT TREATMENT

During the policy debates about housing assistance strategy that took place in the late 1960s, a wide margin of uncertainty about the values of two important policy parameters—the income and price elasticities of demand—became apparent. The available estimates varied a good deal, depending on the nature of the data used in estimation.[1] In order to reduce the uncertainty, the Demand Experiment included "percent of rent" housing allowance plans. The response to the proportional rent rebates provided by these plans permitted econometric estimation of the price elasticity of housing demand. The rent rebates varied the effective price of housing in a controlled way. To facilitate such analysis, recipient households were permitted to choose any rental housing they wished, unconstrained by any housing requirements.

Percent of rent households received a rent rebate, S, equal to a fixed fraction, a, of their gross rent (including utilities), R: S = aR. The household's net housing expenditure, R_n was the difference between their gross expenditure and the rent rebate:

$$R_n = R - S = (1 - a) R. \quad [8.1]$$

Gross expenditures are the product of the quantity of housing services consumed, H, and the price per unit, p_H. The net outlay of the rent rebate recipient was therefore:

$$R_n = (1 - a) R = (1 - a) p_H H. \quad [8.2]$$

Thus, the rent rebate changed the effective relative price of housing from p_H to $(1 - a)p_H$. In the experiment, the subsidy rate a was varied from 0.2 to 0.6 in increments of 0.1, reducing all housing prices to recipients between 20% and 60%. In addition, there was a control group that received no housing price subsidy (a = 0.0). Control households did receive a $10 monthly cooperation payment for providing the same information as experimental households. The assignment of percent of rent households to different subsidy rates was random.

The effect of the rent rebate can be analyzed using a standard microeconomic model of consumer behavior. Assume that households normally consume the quantity of housing services H and nonhousing goods Z that maximize household utility U(H,Z), subject to the budget constraint

$$Y = p_H H + p_Z Z \quad [8.3]$$

where

p_Z = the price on nonhousing goods.

Assume further that the household initially chooses to consume housing of H_0 and nonhousing goods of Z_0.

Under a rent rebate, the household's budget constraint pivots outward as a result of the reduction in the price of housing from p_H to $(1 - a)p_H$ (see Figure 8.1). Desired housing consumption will increase, but nonhousing consumption may decrease, remain the same, or increase, depending on whether the price elasticity of the demand for housing is elastic, unitary elastic, or inelastic, respectively.

The change in housing consumption and the price and income elasticities of demand can be estimated using housing demand functions. In this chapter

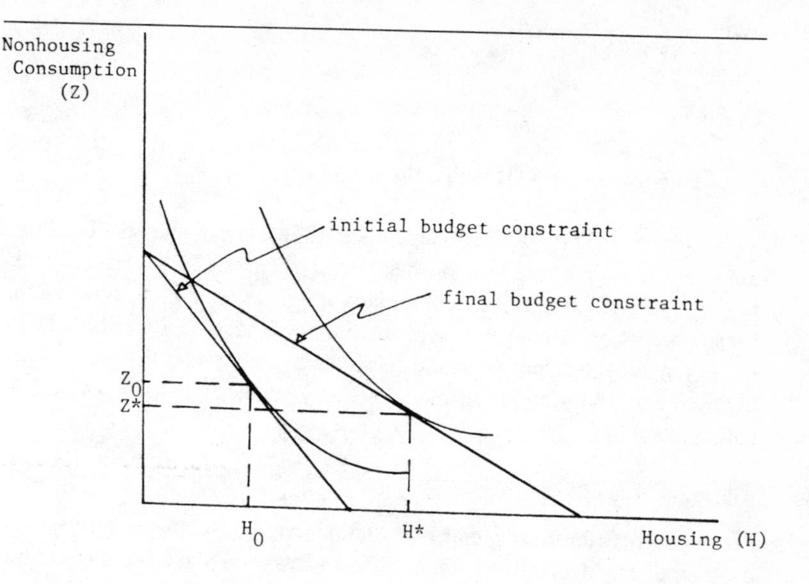

FIGURE 8.1 Effects of a Rent Rebate

we discuss the log-linear demand function which has been widely used in empirical studies of housing demand. This function is written as

$$\ln(H) = k_0 + k_1 \ln(Y) + k_2 \ln(p_H) \quad [8.4]$$

where k_1 and k_2 are the income and price elasticities of demand, respectively. Equation 8.4 can be written in terms of rental expenditures by adding the logarithm of price to both sides. Introducing the percent of rent housing allowances requires, then, modifying both sides of the resultant equation by substituting $(1 - a)p_H$ for p_H and rearranging terms to give

$$\ln(p_H H) = k_0 + k_1 \ln(Y) + (1 + k_2) \ln(p_H) + k_2(1 - a). \quad [8.5]$$

Equation 8.5 contains one unobservable variable, p_H. The equation to be estimated can be rewritten in terms of observable variables and a stochastic error term e as

$$\ln(R) = k_0^* + k_1 \ln(Y) + k_2 \ln(1 - a) + e, \quad [8.6]$$

where

$$k_0^* = k_0 + (1 + k_2)\ln(p_H).$$

As long as Y and $(1 - a)$ are independent of p_H, the parameters of equation 8.6 may be estimated using ordinary least squares (OLS). Experimental households were assigned to rent rebate categories at random, assuring that a was stochastically independent of both income and price. Likewise, there is no reason to believe that income is correlated with the unit price of housing services when housing is considered broadly, specifically to include the location of the unit.

THE ELASTICITY ESTIMATES

The entire sample of renter households may not be the best sample for estimating the demand functions. The housing demand and expenditure functions discussed in the previous section were in theory based on the household's choice of utility-maximizing amounts of housing. The search and moving costs of adjusting housing consumption to changed household circumstances may, however, be significant and therefore may take the household some time. Thus, unless they have moved recently, they may not be consuming their desired amount of housing.[2] Estimates of price and income elasticities based on all households may therefore underestimate the true elasticity, suggesting that a separate demand function should be estimated for movers.

Table 8.1 presents the elasticities estimated for the sample of households moving between enrollment and two years after enrollment. The estimated point elasticities for price are −0.21 in Pittsburgh and −0.22 in Phoenix, while for income they are identical: 0.36 in both sites. The site similarity is striking considering the marked difference in housing market structure between the sites, and it suggests that one demand equation can be estimated for the entire mover sample. When a pooled equation was estimated, the variance ratio test of homogeneity of all regression coefficients (including the constant) across sites indicates rejection of the hypothesis. However, when a site-specific intercept was included to permit housing (and nonhousing) prices to differ between the sites, the variance-ratio test could no longer reject the hypothesis that the elasticities are the same in the two sites. The price elasticity estimated by the pooled log-linear regression is −0.22 and the income elasticity is 0.36, both significantly different from zero at the 0.01 level.

Table 8.1 presents elasticities based on housing expenditure functions that included only price and income as independent variables. Other factors

TABLE 8.1 Price and Income Elasticities of Demand for Housing[a]

	Pittsburgh	Phoenix	Pooled sites
Price elasticity	−0.211	−0.219	−0.216
	(0.063)	(0.059)	(0.043)
Income elasticity	0.363	0.364	0.364
	(0.052)	(0.042)	(0.033)
Sample size	(236)	(292)	(528)

SOURCE: Friedman and Weinberg (1980a: Tables 4.2 and 4.3).
NOTE: Standard errors are in parentheses.
a. All coefficient estimates are significant at the 0.01 level.

TABLE 8.2 Elasticity Estimates for Pooled Sites by Demographic Characteristics

Household Group	*Price Elasticity*	*Income Elasticity*	*Variance-Ratio F statistic*[a]	*Sample Size*
All households	−0.217**	0.366**	0.017	(519)
	(0.044)	(0.033)		
White households	−0.249**	0.413**	0.437	(381)
	(0.048)	(0.036)		
Minority households	−0.183*	0.184**	0.023	(138)
	(0.089)	(0.071)		
Single-person households	−0.175	0.426**	0.723	(65)
	(0.116)	(0.083)		
Single head of household with others	−0.137*	0.294**	0.999	(219)
	(0.063)	(0.058)		
Households headed by a couple	−0.327**	0.468**	1.665	(235)
	(0.070)	(0.061)		

SOURCE: Friedman and Weinberg (1980a: Tables 4.8 and 4.9).
NOTE: Standard errors are in parentheses. Variance-ratio tests indicate that once pooling across sites is performed, pooling across household types is possible ($F(6,507)=1.90$) but pooling across races is not ($F(3,511)=9.17$).

a. Testing site homogeneity allowing site-specific intercepts.
*t statistic significant at the 0.05 level.
**t statistic significant at the 0.01 level.

may also affect housing demand; different demographic groups may have different relative preferences for housing versus other goods and services. Furthermore, policy interest is often focused on certain demographic groups.

The data collected during the Demand Experiment enabled detailed characterization of each household in terms of demographic attributes. A combination of statistical tests, consideration of sample sizes, and intuition was used to reduce the relevant characteristics to two statistically significant and policy-relevant variables: minority status and household composition. Minority status indicated whether the head of the household is a member of a minority group (black in Pittsburgh, black or Hispanic in Phoenix). Household composition indicated whether the household consisted of only a single person (restricted by program rules almost exclusively to elderly persons); was a single head of household (with children or others present); or was a couple (with or without children). These three types of households were denoted single-person, single-headed with others, and couple, respectively.

Pooling the sites was possible for the complete mover sample when demographic differences were ignored (see Table 8.1). Although examination identified different demographic patterns of response in each site, variance-ratio tests could not reject the hypotheses that the two sites could again be pooled when stratified by either race/ethnicity or household composition. Table 8.2 presents these elasticities. Further, once the sites are pooled, homogeneity across demographic groups was rejected for race but not for composition (despite the wide variation in coefficients among the different groups). In sum, it appears that minority households have smaller responses to price and income changes than do nonminority households. (This effect was especially marked in Phoenix.) In addition, while there was evidence of differences in elasticities among household types in Pittsburgh, these differences were not significant for the pooled-site estimates or for Phoenix alone.

THE HOUSING GAP TREATMENT

In contrast to the percent of rent scheme, the housing gap type of allowance payment was constrained, in that it was linked to recipients' housing—households received allowance payments only if their dwelling units met the program's housing standards. The "housing gap" was the difference between the cost of existing housing meeting certain standards (described below) and a fraction of income to be spent on housing. The formula used to determine payments was:

$$S = C - bY \qquad [8.7]$$

where

S = the allowance payment,
C = the basic payment schedule, varied by household size and site,
b = the benefit reduction (household contribution) rate, and
Y = household income.

To provide a contrast with a general income support program, the experiment included an unconstrained plan that offered households a payment based on the housing gap formula (equation 8.7), but without a housing requirement to be met.

The Demand Experiment included 11 different housing gap allowance plans, testing three levels for the basic payment schedule, three values for the benefit reduction rate, and two types of housing requirements—minimum standards and minimum rent. The three basic payment schedules tested were proportional to C*, the estimated cost of existing housing meeting the "minimum standards" for various household sizes in each metropolitan area. The value of the benefit reduction rate, b, varied around 0.25 (corresponding to typical subsidy levels in conventional housing programs).

Households under the minimum standards requirement had to occupy units that met certain physical quality standards for the dwelling unit and had sufficient rooms per person in order to receive payments. Such physical requirements necessitate housing inspections, which are costly to implement and are inconvenient to both tenants and landlords. As a possible less costly alternative, a minimum rent requirement was tested. Minimum rent plans required households to spend at least a certain minimum amount for housing in order to receive allowance payments. Two minimum rent levels were tested—0.7C* and 0.9C*.

There are three possible responses to a housing gap housing allowance offer, illustrated in Figure 8.2. Receipt of an unconstrained allowance payment S moves the budget line outward, inducing the household to consume more housing (H_1). However, a housing gap allowance payment is received only if the household's housing consumption exceeds some minimum amount (H_{min}). The response to the allowance offer depends on the relationship among H_{min}, H_1, and H_0 (the initial consumption level).

In Figure 8.2(a), initial consumption exceeds H_{min} and the household automatically receives the allowance payment. These households simply treat the allowance payment as extra income. Figure 8.2(b) illustrates a second case. This household would not normally meet the housing requirement. If it were to receive the allowance payment, however, the income-induced increase in housing would be sufficient for the household to meet the requirement. Such households, like those in Figure 8.2(a), are in effect unconstrained by the requirement and are free to treat the payment as additional income.

The final case is illustrated by Figure 8.2(c). Households whose housing consumption would be less than H_{min} even with the additional allowance income are constrained to allocate more of the allowance payment to housing than they normally would of income. Because they are required to make what they would consider a nonoptimal allocation, their benefits from the program are lower than their benefits would have been under an unconstrained program. Nevertheless, as long as their utility with the allowance payment and the nonoptimal housing is larger than their utility without the allowance but with their initial housing, they should choose to participate in the program. That is, the household should in theory participate as long as

$$U(H_{min}, Y_0 + S - p_H H_{min}) > U(H_0, Y_0 - p_H H_0). \quad [8.8]$$

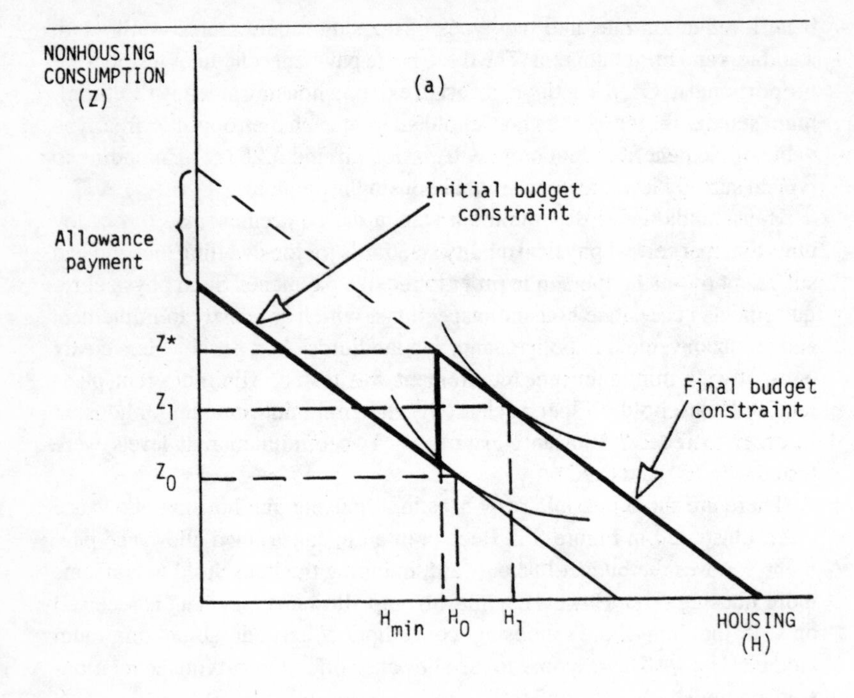

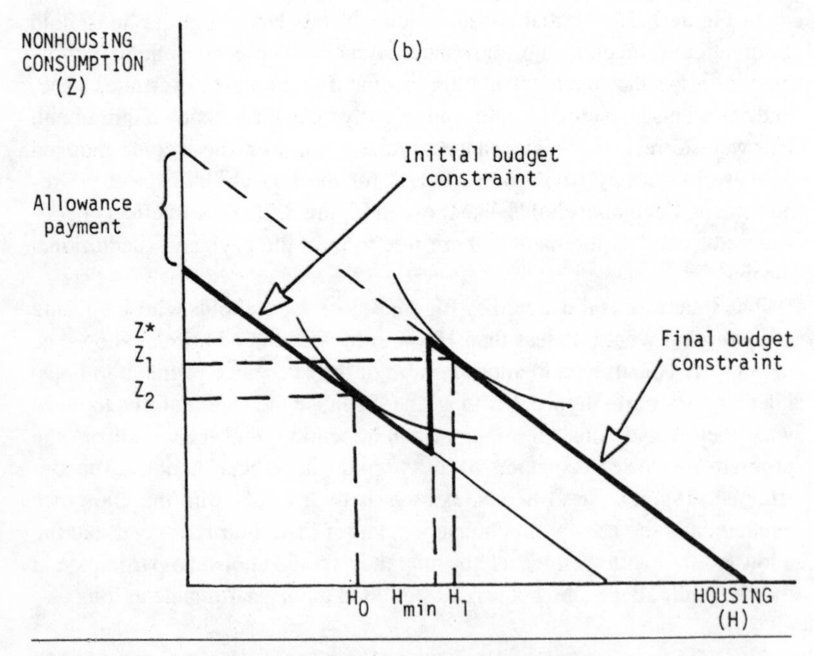

FIGURE 8.2 Allocation of the Allowance Payment to Housing

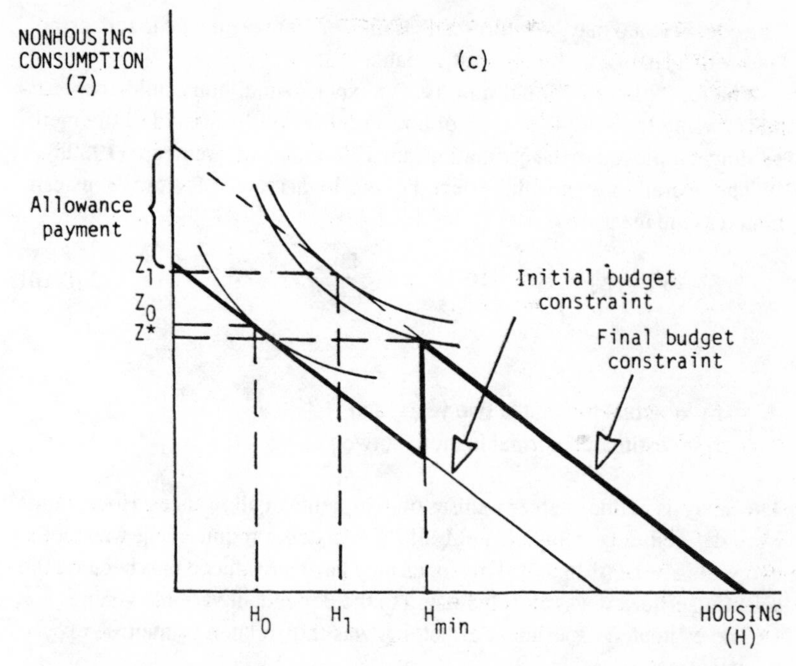

NOTE: The size of the allowance payment relative to income is exaggerated to improve clarity.

FIGURE 8.2 Continued

For some households, however, the payment S will not be large enough to compensate them for their nonoptimal allocation. Such households should not in theory participate in the program.

Experimental effects were measured under the assumption that the actual housing expenditures of households at two years after enrollment, R_A, could be separated into two parts—the normal expenditures that would have been made in the absence of the experiment, R_N, and the additional amount that was induced by the experiment, R_X. Thus $R_A = R_N - R_X$. Because log-linear functions proved useful in analyzing housing demand, the experimental effect was measured in terms of the ratio of actual to normal expenditures (R_A/R_N). Experimental effects were estimated under the assumption that this ratio was functionally related to experimental variables and a random error, specifically

$$\ln(R_A/R_N) = \ln(R_A) - \ln(R_N) = Xd + e \qquad [8.9]$$

where

X = a vector of experimental variables,
d = a vector of experimental effects, and
e = a random error term.

The coefficient d may be interpreted as the percentage change in rent associated with a change in the relevant variable in X.

The logarithm of normal rent, r_N, for experimental households was estimated using the sample of control households. For discussion of the methodology employed in the estimation, see Friedman and Weinberg (1980b).

The overall experimental effect, r_X, the logarithm of R_X, was thus estimated as the mean of

$$\hat{r}_X = r_A - \hat{r}_N \qquad [8.10]$$

where

r_A = actual log rent at two years, and
$\hat{r}_N$ = estimated normal log rent at two years.

The analysis in this chapter focuses on recipients of allowances. Households were defined to be recipients only if their housing requirement was met at two years after enrollment. This focus may have introduced bias because the selection criterion was closely related to the dependent variable—rent.

The estimated experiment effect, $\hat{r}_X$, was thus related to the true experimental effect, r_X, as

$$\hat{r}_X = r_X + w \qquad [8.11]$$

where

w = the expected value of the selection bias for recipients.

Estimation of the bias rested on the tautology that for the entire population (that is, when no subsample of the population is selected), the expected value of the selection bias for the population, e from equation 8.9, was zero. When the entire sample of enrolled households (for which E(e) =0) was divided into three groups, recipient households (status R), households remaining in the sample but not becoming recipients (status $\bar{R}$), and households that dropped out of the experiment before the end of two years (status D), the following relationship was assumed to hold:

$$E(e) = \frac{N_R}{N}E(e|R) + \frac{N_{\bar{R}}}{N}E(e|\bar{R}) + \frac{N_D}{N}E(e|D) = 0, \qquad [8.12]$$

or

$$E(e|R) = -\frac{N_{\bar{R}}}{N_R}E(e|\bar{R}) - \frac{N_D}{N_R}E(e|D) \qquad [8.13]$$

where

N = the total number of enrolled households,
N_i = the total number of households of status i, and
$E(e|i)$ = the expected value of the residual for households with status i.

Under the assumption that $E(e|D) = 0$,[3] the bias w ($w = E(e|R)$) can be determined from

$$w = -\frac{N_{\bar{R}}}{N_R}E(e|\bar{R}). \qquad [8.14]$$

The method used to compute $E(e|\bar{R})$ involved control households whose units did not meet the housing requirement at two years after enrollment.

THE EFFECT ON HOUSING CONSUMPTION

As discussed above, the experimental effect was measured as the percentage of change in housing expenditures above normal rent at two years after enrollment. However, the estimated experimental effect may have to be adjusted for possible bias. For minimum standards households, the estimated selection bias in the expenditures estimate was statistically insignificant and close to zero. In contrast, and as could be expected, the bias in estimates for the minimum rent households was large and significantly different from zero. This was because a minimum rent household's recipiency status was directly related to the household's actual rent outlay, while in the minimum standards plans recipiency was only indirectly related to rent. Direct selection on a dependent variable often leads to bias.

Both the uncorrected and the corrected estimates of the experimental effects on housing expenditures are presented in Table 8.3. The effect for all minimum standards recipient households was statistically significant only in Phoenix, where the increase in expenditures was 14.7% above normal (the effect in Pittsburgh was 3.8%). Separating these households according to their enrollment unit's status with respect to the minimum standards requirement indicates that while the allowance had little or no effect on households living in units that already met the minimum standards at enrollment, it did

TABLE 8.3 Median Percentage Increase in Housing Expenditures Above Normal

	Pittsburgh			Phoenix		
	Median Percentage Change in Expenditures			*Median Percentage Change in Expenditures*		
Household Group	*Uncorrected*	*Corrected*	*Sample Size*	*Uncorrected*	*Corrected*	*Sample Size*
Minimum Standards Recipients	4.3% (2.7)	3.8% (3.2)	84	16.2%** (3.9)	14.7** (4.2)	90
Did not meet requirements at enrollment	7.5* (3.9)	6.5 (4.9)	47	23.6** (5.4)	20.6** (5.8)	63
Met requirements at enrollment	1.1 (3.5)	—	37	−0.7 (3.8)	—	27
Minimum Rent Low Recipients	5.1* (2.6)	2.8 (2.5)	101	19.6** (4.5)	15.7** (4.4)	68
Did not meet requirements at enrollment	17.1** (5.3)	8.7† (5.1)	27	51.5** (9.5)	42.0** (9.3)	26
Met requirements at enrollment	2.4 (2.9)	—	74	−1.2 (3.3)	—	42
Minimum Rent High Recipients	14.9** (3.4)	8.5* (3.6)	57	35.4** (6.0)	28.4** (6.3)	45
Did not meet requirements at enrollment	31.3** (6.1)	15.8* (6.4)	25	53.6** (9.3)	42.6** (9.7)	28
Met requirements at enrollment	4.6 (3.7)	—	32	7.4 (5.0)	—	17
Unconstrained Recipients	2.6 (3.1)	—	59	16.0** (5.6)	—	37

SOURCE: Friedman and Weinberg (1980b: Appendix IX).

NOTE: Standard errors are in parentheses. No correction needed for households that met at enrollment or Unconstrained households.

**t-statistic of estimated effect significant at the 0.01 level.

*t-statistic of estimated effect significant at the 0.05 level.

†t-statistic of estimated effect significant at the 0.10 level.

affect households whose units met the minimum standards only after enrollment. For the group that met minimum standards after enrollment, the median increase in rental expenditures was 6.5% above normal in Pittsburgh and 20.6% above normal in Phoenix, the latter statistically significant.

In Pittsburgh, the minimum rent low plans had only a small effect on expenditures. In contrast, in Phoenix these plans induced rather large and significant increases in expenditures above normal—the median increase was almost 16%. Minimum rent low households that met the requirements only after enrollment had a median increase of 42% above normal, while the change for similar Pittsburgh households was only 9% above normal (significant only at the 0.10 level).

Minimum rent high plans had large and significant effects in both sites, with larger effects, as before, in Phoenix. Minimum rent high plans in Pittsburgh clearly had much larger effects (8% overall and 16% for households meeting after enrollment) than the minimum rent low plans there. In Phoenix, the effects of the two types of minimum rent were similar for households that met the requirements only after enrollment (42% for minimum rent low and 43% for minimum rent high households in Phoenix). Overall, however, the effect of the minimum rent high plans was larger in Phoenix than either the minimum rent low or the minimum standards plans.

The procedure used to estimate the impact of the housing allowance on housing gap households was also used to estimate the impact of the housing allowances on unconstrained households as well. These estimates are presented in the last line of Table 8.3. Only in Phoenix do unconstrained households increase their expenditures significantly more than normal—the increase is only 2.6% above normal in Pittsburgh but is 16.0% in Phoenix. The difference in response between the sites for unconstrained households mirrors the differences for housing gap households.

Since unconstrained households receive a housing gap form of payment without any requirements to meet, comparison can reveal the effect of imposing the requirements above and beyond that of the allowance payment. Table 8.4 presents this comparison for each housing gap group (using the relevant requirement for determination of the initial status for unconstrained households as well). As has been pointed out earlier, housing gap households that already met their requirement at enrollment were essentially unconstrained in their behavior. Thus they would be expected to show the same expenditure changes as the unconstrained group (controlling for payment level). In fact, while Pittsburgh minimum standards households that met requirements at enrollment showed no significant difference in response from unconstrained households, these households in Phoenix increased their housing expenditures significantly less, suggesting that they were reluctant to leave acceptable housing (though there is no evidence that such households moved less often than others; see MacMillan, 1980).

Minimum standards households that met their requirements only after enrollment increased their housing expenditures by more than unconstrained households in both sites. The differences were not large, however, and not significant in either site. As noted elsewhere (Friedman and Weinberg, 1982), however, the minimum standards requirement did induce a significant increase in the probability that a household met those requirements, whereas the unconstrained payment did not. Thus the lack of any differences

TABLE 8.4 Median Percentage Increase in Housing Expenditures for Housing Gap Households Above that for Unconstrained Households (percentage points)

Household Group	*Pittsburgh*	*Phoenix*
Minimum standards recipients	1.1%	−1.1%
	(4.4)	(6.1)
Did not meet requirements at enrollment	2.2	3.6
	(5.8)	(7.8)
Met requirements at enrollment	6.7[a]	−15.2†[a]
	(7.7)	(7.3)
Minimum rent low recipients	0.1	−0.2
	(3.9)	(3.8)
Did not meet requirements at enrollment	6.2	9.6
	(7.2)	(10.9)
Met requirements at enrollment	−1.0	−4.6
	(4.6)	(5.7)
Minimum rent high recipients	5.8†	10.7*
	(3.5)	(5.4)
Did not meet requirements at enrollment	10.5	16.8†
	(7.4)	(10.4)
Met requirements at enrollment	6.1	9.1[a]
	(5.9)	(8.8)

SOURCE: Friedman and Weinberg (1980b: Appendix IX).
NOTE: Standard errors are in parentheses. Housing gap response is corrected for selection bias.
a. Comparison based on 15 or fewer unconstrained household observations.
*t-statistic of the comparison significant at the 0.05 level.
†t-statistic of the comparison significant at the 0.10 level.

in housing expenditure changes may in part reflect the relatively weak link between unit rent and meeting the minimum standards requirements.

The response of minimum rent households can also be compared to that of unconstrained households. Overall, minimum rent low households increased their housing expenditures by about the same percentage as unconstrained households in Phoenix and significantly less in Pittsburgh. Minimum rent high households in both sites increased their expenditures significantly more, though the differences were larger in Phoenix. There was no significant difference in the response of minimum rent households that met their requirement at enrollment from that of comparable unconstrained households.

Minimum rent households that met requirements only after enrollment would be expected to have to spend more on housing than unconstrained households in order to meet the requirements. While some of these households would spend enough to meet the requirements due solely to the income effect of the payment, the requirements were large enough to induce additional expenditures. Only the difference for minimum rent high households in Phoenix was significant, apparently reflecting the relatively small number

TABLE 8.5 Median Percentage Increase in Housing Expenditures for Minimum Rent Households Above that for Minimum Standards Households (percentage points)

	Pittsburgh Percentage Increase		*Phoenix Percentage Increase*	
Household Group	*Minimum Rent Low vs. Minimum Standards Households*	*Minimum Rent High vs. Minimum Standards Households*	*Minimum Rent Low vs. Minimum Standards Households*	*Minimum Rent High vs. Minimum Standards Households*
All Households	−1.0%	4.6%	0.9%	12.0%†
	(3.9)	(4.7)	(5.4)	(6.9)
Did not meet requirements at enrollment	2.0	8.8	17.8*	18.3*
	(6.7)	(7.8)	(9.6)	(9.9)
Met requirements at enrollment	1.3	3.6	−0.5	8.2
	(4.5)	(5.1)	(5.1)	(6.5)

SOURCE: Friedman and Weinberg (1980b: Appendix IX).
NOTE: Standard errors are in parentheses. Housing gap response is corrected for selection bias.
*t-statistic based on estimated contrast significant at the 0.05 level.
†t-statistic based on estimated contrast significant at the 0.10 level.

of unconstrained households available for comparison purposes (and accordingly large standard errors of estimate).

A comparison between minimum rent and minimum standards households is possible as well. Because of the direct link between additional expenditures and meeting the minimum rent requirements, minimum rent households that met requirements after enrollment are likely to have increased their rent by more than minimum standards households. This is generally confirmed by the data in Table 8.5. Minimum rent households that met their requirements after enrollment showed larger increases in expenditures than did minimum standards households that met their requirements after enrollment. The difference is large and significant only in Phoenix, and there was no significant pattern for households that met their requirements at enrollment. For all recipients, minimum rent high households increased expenditures more than minimum standards households (though significantly so only in Phoenix). Minimum rent low households showed the same overall increases as minimum standards households.[4]

POLICY IMPLICATIONS

Housing gap allowance plans are housing programs rather than income maintenance programs, because the payments made to households are linked directly to housing by the imposition of housing requirements. These housing requirements were an important determining factor for household response to the allowance offer. Households that had already met their housing requirements at enrollment, and were therefore automatically eligible for allowance payments, did not use the allowance to pay for any substantial increase in their housing expenditures. Their change in housing consumption was much like what would have occurred in the absence of the experiment. On the other hand, households that met the requirements only after enrollment made large increases in their housing expenditures, well beyond those that would have been made without the program (as estimated using control households).

Similarly to the unconstrained group and households already meeting their requirement at enrollment, percent of rent households used only a modest portion of their allowance payment for additional rent payments. The estimates of the elasticities of housing demand reported here—0.36 and −0.22 for income and price, respectively—are lower than those reported by much of the previous literature. These low elasticity estimates explain the low share of the allowance payment that was used to purchase additional housing (less than one-quarter of the payment went to housing).

These findings illustrate the need to consider the relevant range of estimates whenever making policy recommendations. Thus, for policies concerning the low-income population, such as housing allowances, lower estimates of demand elasticities are probably a more reliable description of their behavior. Even within the low-income population, however, there are differences in the estimated elasticities among demographic groups. If used for projection or simulation, these differences should be taken into account. For policies aimed toward a population with higher income, such as policies dealing with the income tax treatment of home mortgage interest, higher (in absolute value) estimates for the elasticities would probably be more appropriate.

The major problem with a housing assistance strategy based on the housing gap form of housing allowance would appear to be the low participation rate of households in inadequate housing at enrollment. It appears that housing requirements themselves pose a sizable barrier to participation; a fairly large proportion of households living in inadequate housing when they enrolled were still in inadequate housing at the end of two years.

In conclusion, it appears that an unconstrained income transfer can produce more or less the same increases in housing consumption as a constrained transfer. However, if policymakers are concerned with improving specific facets of housing consumption, this is best achieved by imposing exactly those physical requirements. Such a housing allowance strategy can, however, markedly reduce the number of recipients.

NOTES

1. See Mayo (1981) for a review of the housing demand literature.

2. See Weinberg et al. (1981) for a discussion of a microeconomic disequilibrium model of residential mobility.

3. Theoretically, there is no clear indication of the size or sign of $E(e \mid D)$. Voluntary dropouts consist of two groups—those whose housing consumption is so small that they had little expectation of meeting the requirements, and those whose incomes rose so much that their allowance payment would be negligible. The former group has below-average rents ($e < 0$), while the latter has above-average rents ($e > 0$). In an attempt to determine the empirical importance of $E(e \mid D)$, the relationship of the enrollment rent of dropouts to that of all households at enrollment was determined (due to high serial correlation, this is a good measure of the expected residual at two years). The enrollment residual was small and insignificant.

4. For a discussion of the site differences in response, see Friedman and Weinberg (1982).

9

The Use of Hedonic Indices

SALLY R. MERRILL

□ ALTHOUGH HEDONIC INDICES have been widely applied in analyses of housing and urban issues, the housing allowance experiments have probably provided the richest opportunities to date to use hedonic indices in innovative and important ways. Hedonic indices are used in two major ways in the experiments: to analyze experimental effects and to explore a variety of housing market analyses.

Hedonic indices essentially attempt to sort out the influence of housing and nonhousing factors in determining the market value of units. This allows the construction of indices that are sensitive to both unit and neighborhood characteristics and that do not include nonhousing factors such as inflation, tenure conditions, or racial discrimination. Empirically, the hedonic approach refers to the systematic regressions of rent on housing characteristics. The hedonic model estimated for housing is generally of the form

$$R = F(D,N,A,T,H) \quad [9.1]$$

where rent (R) is expressed as a function of dwelling unit characteristics (D), neighborhood attributes and amenities (N), access characteristics (A), and conditions of tenure (T). Frequently, household characteristics (H), such as race, are also included in order to test for price differentials based on these characteristics.

The hedonic quality index used in the analysis below is constructed as follows. Assuming that Q_i is the vector of housing quality attributes for a particular unit (that is, D, N, and A) and NQ_i is the vector of nonquality attributes (that is, T, H, and possibly other factors such as market segmentation), one can estimate

$$R_i = F(Q_i, NQ_i, e_i) \quad [9.2]$$

where e_i is stochastic. One then uses $\hat{R}_i = F(Q_i, \overline{NQ_i}, O)$ as an index of housing quality, where $\overline{NQ_i}$ represents some fixed values for the nonquality items.

One of the major themes of Demand Experiment research was a parallel analysis of experimental outcomes using both expenditures and the hedonic value of housing services. The key question being addressed was whether differences between changes in expenditures under HADE and changes in hedonic housing quality followed the normal pattern of rent-hedonic quality differences. If they did, we can conclude that expenditure changes induced by the experiment led to concomitant changes in real housing services. The results strongly suggest that experimental households frequently did *not* obtain the level of housing services that normally accompanies increased expenditures. Indeed, the discrepancy initiated a lengthy investigation of the hedonic residuals to assess search behavior, experimentally induced shopping inefficiency, and the appropriateness of the specification of the hedonic equation. The results suggest that there are price differences (there are "good" and "bad" deals for housing), that the shopping behavior of households in response to these differences will respond to changes in the incentives to shop more or less extensively, and that shopping effects can be analyzed successfully using hedonic indices.

Other analyses using hedonic indices included an assessment of quality change for experimental households that did not move (maintenance and upgrading); a comparison of housing outcomes and cost-effectiveness of housing allowances and other rental housing assistance programs in HADE sites; tests for racial price discrimination and other types of market segmentation; evaluation of the "portability" of indices across cities; and estimation of demand functions for housing attributes. Because of differences in experimental design and analytical emphasis, hedonic indices were much more widely used in the Demand Experiment than in the Supply Experiment.

In many hedonic analyses, the ability of the available data to meet the demands imposed by hedonic index construction has been a serious problem. Fortunately, extensive data on housing neighborhood, location, and tenure attributes were collected for the Housing Allowance Experiments. Thus, a great deal of attention was devoted to specification in both HADE and HASE. Development of the indices is discussed at length elsewhere (for HADE see Merrill, 1980; for HASE see Barnett, 1979, on Brown County and Noland, 1980, on St. Joseph County).[1]

The discussion below focuses on two of the most important applications of hedonic indices in the experiments: the analysis of experimental effect in the Demand Experiment, including an evaluation of the hedonic residuals, and estimation of component demand functions using HASE and HADE data.

A COMPARISON OF EXPERIMENTAL OUTCOMES IN HADE

Experimental impact for housing gap and percent of rent households was estimated using both housing expenditures and the hedonic measure of housing services. The goal was to isolate real changes in housing services in response to the experiment. Thus, changes in rent would be allocated between changes in housing quality and price changes due to inflation, changes in length of tenure, changes in landlord characteristics, and more or less effective shopping behavior. The methodology used to determine impact was essentially the same for both measures. A detailed presentation of the methodology and the results are discussed at length in Friedman and Weinberg (1980a, 1980b), and is also summarized in this volume (see Chapter 8).

Tables 9.1 and 9.2 summarize some of the key results for housing gap and percent of rent households for both expenditures and hedonic quality, respectively. Clearly, an increase in housing expenditures does not imply a commensurate increase in real housing consumption. The experimental effects for housing services are consistently less than those for expenditures in both Pittsburgh and Phoenix. Real housing consumption increased less than expenditures for both minimum standards and minimum rent households. No experimental effect for housing services was observed in Pittsburgh; although the estimated effect was significant in Phoenix, it was consistently much smaller than for expenditures.

The most striking difference occurred for the percent of rent allowance payment (see Table 9.2). Income elasticity estimates for both expenditures

TABLE 9.1 Estimated Percentage Increase Above Normal in Housing Expenditures and Hedonic Housing Services (HADE) (standard errors in parentheses)

Household Group / *Outcome Measure/Site*	*Households that Did Not Meet Requirements at Enrollment But Met at Two Years*		
	Minimum Standards	*Minimum Rent Low*	*Minimum Rent High*
Pittsburgh			
Housing expenditures	7.5*	8.7†	15.8**
	(3.9)	(5.1)	(6.4)
Sample size	47	27	25
Housing services	5.6	−0.9	3.1
	(4.1)	(4.4)	(4.8)
Sample size	43	20	23
Phoenix			
Housing expenditures	23.6**	42.0**	42.6**
	(5.4)	(9.3)	(9.7)
Sample size	63	26	28
Housing services	10.5*	20.2**	26.0**
	(4.7)	(7.2)	(7.3)
Sample size	50	20	25

SOURCE: Friedman and Weinberg (1980b: Tables 5.1, 5.10, 5.11, 6.7, 6.9, 6.10).
*t-statistic of estimated effect significant at 0.01 level.
**t-statistic significant at 0.05 level.
†t-statistic significant at 0.10 level.

and housing services are significant and reasonably large. Surprising differences are evident in the price elasticity estimates, however. In Pittsburgh, the estimated price elasticity for housing services is only about half that for housing; in Phoenix, the estimated price elasticity for housing services is not even significant. Another aspect of the results is the marked discrepancy in outcomes for minority and nonminority households. Minority households have an insignificant (and positive) estimated price elasticity for housing services in both Pittsburgh and Phoenix, while this is not the case for nonminority households.

One explanation for these results is experimentally induced inefficiencies in search and shopping behavior. The arguments are as follows. A rent rebate program such as a percent of rent allowance that provides no direct incentive for households to increase their housing quality may lead to inefficiency in shopping behavior. A household with a 50% rebate, for example, would pay

TABLE 9.2 Price and Income Elasticities for Expenditures and Hedonic Housing Services (HADE) (standard errors in parentheses)

	Pittsburgh	*Phoenix*
Price Elasticity		
Expenditures estimate	−.230**	−.215**
	(.065)	(.064)
Nonminority	−.233**	−.290**
	(.070)	(.073)
Minority	−.207	−.154
	(.181)	(.115)
Housing services estimate	−.113*	−.045
	(.054)	(.060)
Nonminority	−.143*	−.129*
	(.057)	(.065)
Minority	.067	.023
	(.201)	(.106)
Income Elasticity		
Expenditures estimate	.338**	.353**
	(.054)	(.046)
Nonminority	.358**	.406**
	(.059)	(.050)
Minority	.232	.169*
	(.136)	(.091)
Housing services estimate	.226**	.375**
	(.047)	(.043)
Nonminority	.269**	.440**
	(.048)	(.045)
Minority	−.012	.154
	(.151)	(.085)

SOURCE: Friedman and Weinberg (1980a: Tables 5.4, 5.8, X.41).
*t-statistic of estimated effect significant at 0.01 level.
**t-statistic significant at 0.05 level.
†t-statistic significant at 0.10 level.

only half of any overpayment (relative to average market costs for such a unit) and likewise gain only half of any reduced payment achieved by more careful shopping. If households no longer pay full market price for each additional unit of housing services, the rent subsidy may permit them to reduce search costs. Similarly, the allowance offer could have altered the shopping behavior of households in the minimum rent plans. If at enrollment, for example, the household spent too little money on rent to pass the requirement, it would have to find a more expensive unit to receive an allowance. This could lead the household to prefer a unit that normally would be considered a "bad deal" (the unit price of housing services exceeds the market average) but which passed the minimum rent requirement over another unit that was a "good deal" but did not pass minimum rent. For minimum standards households the argument is essentially similar, although the effects would probably be less direct because of the incentive to find units that passed minimum standards. Presumably, minimum standards households might stop shopping earlier than they normally would after finding a unit that met requirements.

ANALYSIS OF THE HEDONIC RESIDUALS

The critical issue in understanding these findings is whether lower estimated price elasticities based on the estimated hedonic values really reflect induced inefficiencies in shopping behavior or something else. Addressing this issue involves analysis of the hedonic residual (the difference between rent and predicted rent). In theory, the hedonic residual reflects price differences for units offering equivalent housing services (Kennedy and Merrill, 1979). In practice, however, the hedonic index may be subject to several types of specification bias. For example, if important attributes of the housing bundle were omitted from the estimating equation, the index would not adequately reflect the unit's housing services. Omitted variables increase the estimated standard error of the hedonic index. If important positively valued attributes of dwelling unit or neighborhood quality are missing from the hedonic estimates, the estimated price and income elasticities for housing services would be biased downward from the true elasticities.

Next, if the housing market in Pittsburgh or Phoenix is segmented—that is, if different groups of households (central-city versus suburban or racial differences, for example) face different housing prices—the same set of relative attribute prices estimated by an overall index may not be applicable to all submarkets. Because of the apparent differences in outcomes for minorities and nonminorities, market segmentation was carefully assessed; separate hedonic indices were estimated for racial submarkets, but the results were not changed (see Merrill 1980; Friedman and Weinberg, 1980a).

The analysis suggests that, in fact, the residuals reflect both shopping behavior and omitted variables. (The analyses summarized below are presented in Kennedy and Merrill, 1979; Merrill, 1980; and Friedman and Weinberg, 1980a.) Based on the estimated effects of the omitted variables, the price and income elasticity estimates can be corrected for the effects of omitted variables. Formally, the omitted variable problem can be expressed by dividing the dwelling unit characteristics in the hedonic equation (equation 9.3) into two categories—those included in the hedonic equation (X) and those omitted (Z). Thus,

$$\ln R = X\beta + Z\lambda + T\tau + \theta \qquad [9.3]$$

where

R = a vector of unit rents;

X = a matrix of unit, neighborhood, and location characteristics included in the hedonic equation;

Z = a matrix of (relevant) unit characteristics not included in the hedonic equation;

T = a matrix of tenure conditions;

β,λ,τ = unknown coefficients; and

θ = a vector of stochastic terms reflecting pure price effects.

Notice that all relevant housing services (including higher-order terms in any variables) appear in the equation by definition. Thus, the stochastic term reflects only differences in the amount paid for a given type of housing. Positive values of θ mean that the household has a bad deal, negative values mean that it has a good deal.

Equation 9.3 divides the hedonic error term into two components—one due to omitted variables and the other reflecting unit price differences (good or bad deals). Estimates of the hedonic equation based on X alone are biased. The estimates also reflect the correlation of included items with the omitted variables. The estimated residual, $\hat{\theta}$, thus reflects the effects of both omitted items and price effects, net of the effect of sampling error on the estimates (for derivation, see Kennedy and Merrill, 1979).[2]

The proxy relation between the omitted variables and included variables may simply be an empirical characteristic of the observed sample. There is some reason to believe, however, that the proxy relationships may instead reflect underlying behavior and thus may be reasonably stable. The omitted variable problem does not arise solely because of inadequate data. In fact, housing characteristics are highly correlated. The Demand Experiment collected much more extensive data on housing conditions than is usually available. Even so, the final equations involved only 24-34 variables depending on the site and exact specification and, despite considerable exploration of alternative forms, additional variables, higher-order terms and interactions, and almost 1600 observations in each site. Many of the original variables were highly correlated and had to be either omitted entirely or aggregated (using principal components and other data reduction techniques). The presumption is that these strong correlations reflect demand and production and are not simply chance patterns of association.

This is not to say that all omitted items are highly correlated with the included items. Still, the overall explanatory power of the estimated regression is good. Equations were estimated for each site using both a linear and semilog specification. The R^2 for the two specifications is reasonably high—0.66 in Pittsburgh and 0.80 in Phoenix.[3] The estimated residual values are not trivial, however: They have an estimated standard deviation of about 17% of the mean value of the dependent variable for the linear specification and 4% for the semilog specification. Their range in the sample was roughly ±60% of the mean value of the dependent variable for the linear specification and ±13% for the semilog specification.

Several hypotheses can be tested to determine the correct interpretation of the estimated residual, $\hat{e}$. If the residual involves some omitted quality, then it should be positively correlated with household income and household satisfaction. If the residual reflects changes in shopping behavior, then the search behavior of percent of rent households should show some differences from control households.

First, it seems reasonable on the face of it to assume that expressed satisfaction will generally increase with the hedonic value of the unit (not including tenure variables). This is indeed the case. Regression of pre-experimental expressed satisfaction with dwelling unit and neighborhood on a variety of demographic characteristics, the hedonic value, the tenure terms, and the hedonic residual shows a strong positive association between both dwelling unit and neighborhood satisfaction and the hedonic value of the unit.

Second, if the hedonic residual includes some omitted quality (i.e., quality unmeasured by the hedonic index), then to the extent that satisfaction is positively related to the level of housing quality, the residual should be positively related to the household's satisfaction with its dwelling unit. If, on the other hand, the hedonic residual is due largely to price effects rather than omitted quality, the association with satisfaction is expected to be negative—that is, that satisfaction increases as the amount of quality relative to expenditures increases. The change in hedonic quality and the change in the hedonic residual over the two years of the experiment were each regressed on the change in dwelling unit satisfaction for control households. The results showed that the changes in quality and satisfaction have a significant and positive relationship in both Pittsburgh and Phoenix. Further, in both sites, the change in satisfaction and the change in the hedonic residual have the expected significant and negative relationship. Thus, it seems that the hedonic residual is not solely due to omitted quality items.

Similar hypotheses were tested concerning the relationship of search effort to quality and price effects. If the hedonic residual reflects price heterogeneity, then a diligent search will result in a better deal (more quality per dollar) than would a haphazard search. If true, the hedonic residual will be negatively associated with search effort. One measure of the search effort is the number of days spent searching for housing. As shown in Kennedy and Merrill (1979: Table 6), increased search time does result in getting a better deal in both Pittsburgh and Phoenix. Search time is also associated with a larger change in the quality index in Pittsburgh.

Part of the smaller increase in housing services relative to expenditures for percent of rent households in both Phoenix and Pittsburgh may be due to a conscious decision on their part to use less effort in searching for a new unit. Since there is a significant association between increased search time and obtaining more housing services per dollar of expenditures, if percent of rent households search less than control households, then the price discount will have a smaller effect on their housing services than on their expenditures. There is some weak evidence that decrease in search effort occurred—percent of rent movers spent fewer days looking for a unit than did control movers, though not significantly so: 97 versus 119 days in Pittsburgh; 34 versus 46 days in Phoenix (see Friedman and Weinberg, 1980a: Appendix Table X.38).

Another approach to analyzing the residual is direct estimation of demand for the residual. During the development of the hedonic index, extensive analysis of the residual was carried out using the entire enrolled sample (Merrill, 1980). The hedonic residuals and the percentage of deviation of predicted and actual rent were regressed on household income, race, household size, and age and education of head of the household. The major hypothesis tested was the following: If important quality attributes were omitted, there would be a significant positive relationship between the residual and income and perhaps education. The income coefficients were in fact significant but were extremely small in both Pittsburgh and Phoenix.

A series of similar models have been estimated for households remaining active in the experiment for two years. Simplifying, equation 9.3 becomes

$$\ln(R) = \ln(Q) + \ln(T) + \ln(\epsilon) \qquad [9.4]$$

where

TABLE 9.3 Price and Income Elasticity Estimates for Rent Components (HADE)

	Pittsburgh		*Phoenix*	
Dependent Variable	*Price Elasticity*	*Income Elasticity*	*Price Elasticity*	*Income Elasticity*
Rent	−0.230**	0.338**	−0.215**	0.353**
	(0.065)	(0.054)	(0.064)	(0.046)
Hedonic index	−0.113*	0.226**	−0.045	0.375**
	(0.057)	(0.047)	(0.060)	(0.043)
Hedonic residual	−0.159**	0.089*	−0.193**	−0.021
	(0.047)	(0.039)	(0.048)	(0.034)
Tenure characteristics	0.027*	0.019†	0.017	0.001
	(0.013)	(0.010)	(0.011)	(0.008)
Definition difference[a]	0.016	0.004	0.005	−0.002
	(0.013)	(0.010)	(0.007)	(0.005)
"Corrected" hedonic index	−0.158	—	−0.045[c]	—
price elasticities	(0.082)	—	(0.170)	—
"Shopping effect"[b]	−0.072	—	−0.170	—
Sample Size	(214)		(257)	

SOURCES: Friedman and Weinberg (1980a: Table 5.5) and Kennedy and Merrill (1979: Table 8).

a. Difference between the analytic rent variable used for the expenditure analysis and that used in the derivation of the hedonic index.

b. Difference between the expenditures price elasticity and the "corrected" hedonic index price elasticity.

c. Corrected price elasticities are computed as indicated in equation 9.6 with the following exceptions. First, the correction applies only in theory when the income elasticity of the residual is positive (when the hedonic income elasticity is less than the expenditure income elasticity). Cases other than this are assumed to represent stochastic error, with a true value of η_y^ϵ of zero (so that the hedonic price elasticity is not changed).

**t-statistic significant at 0.01 level.

*t-statistic significant at 0.05 level.

†t-statistic significant at 0.10 level.

$\ln(Q) = (\alpha + X\beta)$, the hedonic index of housing services (unit and neighborhood characteristics included in the equation) abstracting from tenure characteristics;

$\ln(T) = T\tau$, the value of tenure characteristics; and

$\ln(\epsilon) = \hat{\theta}$ the estimated residuals.

If each component of equation 9.4 [$\ln(R)$, $\ln(Q)$, $\ln(T)$, and $\ln(\epsilon)$] is regressed on the logarithms of price and income, it will be true that equation 9.4 implies the sum of the price (or income) elasticities for Q, T, and ϵ will equal the price (income) elasticity of R (see Friedman and Weinberg, 1980a).

These elasticities were estimated using log-linear regressions of the residual, the quality index, tenure adjustments, and rent adjustments on price and income and are summarized in Table 9.3. In Pittsburgh, both the price and income elasticities of housing services fall relative to those for expenditures.

The difference is accounted for almost entirely by the hedonic residual and not the tenure characteristics. Since there is a significant positive income elasticity for the residual in Pittsburgh, it becomes plausible to assume that the residual in that site at least partly represents omitted quality variables. The presence of a significant price elasticity for the residual in both sites suggests that price factors are also present in the residual.

The ratio of the elasticities can also provide some information on the relative importance of omitted quality and price effects. The significant income elasticity in Pittsburgh suggests that, in that site at least, the hedonic residual does include some omitted quality. In this case, the estimated price and income elasticities based on the hedonic index would underestimate the true elasticities of housing services; at least part of the change in the hedonic residual would represent real changes in housing in addition to the changes reflected by the index.

One way to correct this problem may be to use the income elasticity of the hedonic residual. If this represents increased expenditure for housing services not included in the hedonic index, then households allocate increased expenditures between included and omitted items in the proportion $(\eta_y^\epsilon/\eta_y^Q)$, where η_y^ϵ and η_y^Q are the income elasticities of the hedonic residual and hedonic index, respectively. Since the rent rebates offered by percent of rent apply to both included and omitted quality, it seems reasonable to suppose that the allocation of increased expenditures between the items included in the hedonic index and those included in the hedonic residual would be the same as that for increases arising from higher incomes (Kennedy and Merrill, 1979). One can then write the following:

$$\frac{\tilde{\eta}_p^\epsilon}{\eta_p^Q} = \frac{\eta_y^\epsilon}{\eta_y^Q} \qquad [9.5]$$

where

$\tilde{\eta}_p^\epsilon$ = the (unknown) price elasticity of omitted items; and

η_p^Q = the price elasticity of housing services included in the hedonic index.

In other words, the ratio of the price elasticities for omitted items to that for housing services is assumed to be the same as the ratio of the income elasticities, where the income elasticity for omitted quality is approximated by the income elasticity for the hedonic residual. Equation 9.5 can be solved to give the estimated price elasticity of omitted items as a function of the estimated hedonic index price and income elasticities and the estimated hedonic residual income elasticity:

$$\widetilde{\eta}_{p}^{\epsilon} = \frac{\eta_{p}^{Q} \cdot \eta_{y}^{\epsilon}}{\eta_{y}^{Q}}. \qquad [9.6]$$

Put another way, equations 9.5 and 9.6 essentially accept the income elasticity of expenditures as a benchmark for the normal relation between changes in expenditures and changes in the hedonic index (ignoring tenure characteristics and definitional differences, which have only a small effect). The adjusted price elasticity ($\widetilde{\eta}_{p}^{\epsilon}$) incorporates this normal relation. The difference between the expenditure price elasticity and the adjusted price elasticity thus measures the extent to which rent rebates altered the normal relationship between housing expenditures and housing services. It is worth emphasizing the arbitrary nature of this procedure. Other methods of correction are possible.

Table 9.3 implies a value for $\widetilde{\eta}_{p}^{\epsilon}$ in Pittsburgh of −0.045. Since the total price elasticity of housing services is the sum of the hedonic elasticity and the elasticity of omitted items, this implies a "corrected" hedonic price elasticity in Pittsburgh of −0.158. Since the expenditures price elasticity is −0.230, this implies a shopping effect of −0.072—that is, only about two-thirds of the expenditure increase induced by the percent of rent plans goes to increased housing services. In Phoenix, the income elasticity is small, negative, and not significantly different from zero. Furthermore, the model behind equation 9.5 considers only the possibility of a positive income elasticity for the residual (reflecting omitted quality). Therefore, no adjustment in the Phoenix housing services elasticity is indicated, and the "shopping effect" remains large.

In conclusion, the evidence in Phoenix seems most consistent with viewing the hedonic residual solely as a price effect, representing changes in shopping behavior, not as omitted quality. The income elasticity estimates are the same for the quality index and rent and zero for the residual. Almost all of the difference in price elasticity estimates is found in the residual. On the other hand, the residual in Pittsburgh represents both some omitted quality and some price behavior changes. Thus, the hedonic index price elasticity was "corrected" in Pittsburgh (but not in Phoenix); the resultant shopping effect of the percent of rent price subsidy was moderate in Pittsburgh and substantial in Phoenix.

COMPONENT DEMAND FUNCTIONS

A number of researchers have explored attribute demand functions (for example, King, 1976; Kain and Quigley, 1975; Strazheim, 1975) in order to assess the effects on consumption patterns of demographic characteristics, attribute prices, racial segregation, and so forth. Both HASE and HADE data were used to estimate component demand functions: Barnett (1981) specified demand functions for two summary attributes (space and quality), while

Merrill (1976) estimated functions for numerous individual attributes, primarily to test for racial differences in consumption.

In Barnett and Noland's (1981) model, households' demand for space and quality depends on how much they spend for housing, the prices of space and quality, and taste. Thus:

$$S = f(E/P_q, P_s, T) \qquad [9.7]$$

$$Q = g(E/P_q, P_s, T) \qquad [9.8]$$

where

S = the quantity of space demanded;
E/P_q = housing expense normalized by the price of quality;
P_s = the price of space;
T = taste; and
Q = the amount of quality demanded.

The functional form used by Barnett to estimate the parameters of the demand function was:

$$S = \beta_o \cdot \prod_{i=1}^{5} x_i \beta_i \cdot \exp \sum_{j=1}^{4} d_j \delta_j \qquad [9.9]$$

where

β_o = the constant term;
x_1 = housing expense (E in equations 9.7 and 9.8);
x_2 = number of persons in household;
x_3 = number of minors;
x_4 = age of head of household;
x_5 = education of head of household;
β_i's = elasticities;
d_1 = indicator variable for households containing a husband and wife;
d_2 = indicator variable for households headed by a single parent;
d_3 = indicator variable for nonwhite households;
d_4 = indicator variable for households living in St. Joseph County;
δ_j = percent change in space consumption associated with being in the indicated group.

Barnett's model rests on several assumptions. First, in choosing among units, consumers consider the mix of attributes rather than just the overall "amount" of housing services. Second, households choose their optimal consumption bundle in two steps, first deciding how much to spend on housing and then allocating the total across housing attributes. Third, every consumer faces the same set of attribute prices. Since this was verified empirically (that is, attribute prices do not differ significantly for renters and owners in the two HASE sites), equation 9.9 does not include prices. Finally, Barnett assumes that dwelling units consist of only two attributes, space and quality. Based on the hedonic index for Brown County, space (S) is measured as $S = \$48 \times \ln(\text{rooms}) + \$19 \times (\text{bathrooms})$. Thus, quality (Q) is simply expenditures (E) less space: $Q = E - S$. Although Barnett recognizes that this definition of quality, which includes location, could distort inferences if the demands for quality and location differ, he argues that the HASE sites are relatively small and location attributes contribute very little to variation in rent (Barnett, 1979; Noland, 1980). It is not necessary to fit an equation like (9.9) for quality because, by definition, expenditures for space and quality exhaust housing expense.

To avoid the bias that might arise using OLS (see Barnett, 1981: 8), a two-stage estimation was used. The first stage regressed housing expense on the logarithm of households' current income and all the independent variables in equation 9.9 except housing expense. The second stage substitutes predicted housing expense from the first stage for actual housing expense and then estimates equation 9.8 with OLS. The demand equations were fit separately for owners and renters because the data indicated that their demand equations were statistically distinct. The results are presented in Table 9.4. The implied elasticities for quality are derived by using the fact that the elasticity of space weighted by its share of housing expense and the elasticity of quality weighted by its share of expense must total 1. The estimated and implied elasticities are:

	Renters	*Owners*
Space	0.26	0.35
Quality	2.32	1.65

The difference between elasticities of the two attributes is striking. A 10% increase in housing expense will increase the demand for space by only about 3% while the demand for quality will go up by 17%-23%, depending on tenure.

Although housing expenses clearly influence the amount of space demanded, household size and structure are more important determinants, judging by their t-values. For example, a 100% increase in household size will increase the demand for space by 10%-20%. The effect of compositional differences among households varies by tenure: In a household of a given size, the number of children in the household affects renters' demand but not owners'. A renter household with three adults and one child consumes, on average, about 3% more space than an otherwise comparable household with two adults and two children. That difference seems sensible: Adults need more private space. However, a difference in the mix of adults

TABLE 9.4 Component Demand Functions, Renters' and Owners' Demand for Space (HASE)

Independent Variables	*Renters*		*Owners*	
	Coefficient	*t-statistic*	*Coefficient*	*t-statistic*
Housing expense (1n)	.255**	4.67	.350**	7.015
Number in household (1n)	.195**	11.164	.102**	5.888
Number of children (1n)	−.039**	3.203	.000	.015
Couples	−.031**	3.399	−.085**	5.789
Single parent	.014	.969	−.082**	2.985
St. Joseph County	.002	.282	.049	3.939
Age (1n)	.078**	8.316	.060**	4.340
Education (1n)	−.021	1.224	.023	1.411
Black	.020	1.246	.088**	3.214
Constant	2.847**	11.802	2.415**	9.482
	R^2=0.34		R^2=0.23	
	F=140.00		F=33.93	
	Sample Size=2,512		Sample Size=1,056	

SOURCE: Barnett and Noland (1981: Tables 3 and 4).
NOTES: The dependent variable is the natural log of space. The equations are fit with 1974 data from Brown County and 1975 data from St. Joseph County.
**t-statistic significant at 0.01 level.
*t-statistic significant at 0.05 level.
†t-statistic significant at 0.10 level.

and children in owner households does not affect their demand for space. Renters and owners thus appear to respond differently to changes in their household structure: Renters tailor their consumption; owners do not.

Interestingly, Barnett hypothesizes that this asymmetry may be due to renter/owner differences in closeness to equilibrium. It costs renters less to move than owners, since renters do not have to pay the substantial transaction costs associated with selling one home and buying another. Renters' consumption of space and quality should therefore be closer than owners' to their equilibrium demand. As a consequence, owners' measured household size may bear an errors-in-variables relationship to their consumption. Their current housing may have been chosen for a household larger or smaller than they currently need; that is, either in anticipation of a larger household size in the future (e.g., young couples) or to accommodate a larger household in the past (e.g., elderly households). Alternatively, they may not have anticipated their current size.

Barnett summarizes his key findings as follows:

- Space is a necessity that households buy first. As their housing expenses increase, a larger fraction goes to buy more quality.
- Household size and structure are important determinants of space and quality demand. For example, larger families and older families spend more of their housing dollar on space; couples spend comparatively less.

- Even though owners and renters have statistically distinct demand functions, differences in their consumption can be almost completely accounted for by the owners being older, having larger families, and spending considerably more. Almost none of the difference in consumption is attributable to the differences between owners' and renters' demand function.

SEGREGATION AND HOUSING QUALITY: PITTSBURGH AND PHOENIX

During the development of the hedonic indices for the Demand Experiment, a number of tests for market segmentation were carried out, including those for price discrimination, on the basis of race of household and submarkets of different racial composition. Residents of ghetto areas in Pittsburgh—that is, in submarkets where more than 50% of the residents are black—appear to pay a price premium. It is small, however—about 4%. No evidence of price discrimination against either black or Hispanic households was found in Phoenix (Merrill, 1980).

An additional assessment of the effects of segregation on the type and quality of housing consumed was carried out using attribute demand functions (see Merrill, 1976). The hypothesis tested was the following: If marked differences exist in the attributes of ghetto and nonghetto housing and in the price of these attributes, then the consumption patterns of residents of these submarkets will differ. And if discrimination limits the locational choices of minority households, then minority and white households having the *same* income and other socioeconomic characteristics may consume very different housing. Many minority households are constrained to live in particular submarkets; as a result, they cannot "reveal" their preferences in the overall market in the same manner as whites. Distortions caused by segregation may be particularly severe for moderate-income minority households. For example, on the basis of their income, family characteristics, and work site, more of these households would be expected to live in the suburbs than is actually the case (King and Mieszkowski, 1973).

In order to test for systematic racial differences in housing consumption patterns, consumption functions were estimated for individual housing attributes, including most of the important attributes used in specifying the hedonic equations. The independent variables include demographic factors expected to influence the consumption of these attributes—income, level of education, age, family size, employment status, occupational status, and household composition.

The effects of race on consumption were tested in three ways. First, dummy variables indicating minority head of household were included in equations estimated for each city. The coefficient of the minority household variable measures the average effect on consumption of that attribute, for all minority households in the sample. However, since the distorting effects are presumed to be most severe in ghetto areas, dummy variables indicating residence in the minority or racially mixed submarkets were to be used in a separate estimation. The coefficients of these submarket variables measure the average differences in attribute consumption for residents of ghetto and mixed markets relative to residents of the white submarket. One additional factor must be considered, however. Expenditure functions estimated for the full enrollment sample indicated that minority households spend less on rent at the same level in income. On average, blacks in Pittsburgh spend about 10% less than white households; in Phoenix, Hispanic households spent about 19% less. Controlling for income may therefore be inadequate to isolate potential distortions in consumption patterns caused by segregation. Although unlikely, it is conceivable that minority and white consumption patterns are the same but that minority households simply consume proportionately less of each attribute at a given income level. Therefore, in order to better assess whether segregation distorts minority housing consumption, it is relevant to ask whether minorities paying the same *rent* as whites consume different amounts of attributes. In order to address this question, component demand functions were estimated controlling for rent level rather than for income.

Several cautions must be expressed concerning the econometric approach to estimating the attribute demand functions. Two types of equations are estimated. The quantity of an attribute purchased is presumed to be a function of either income and household characteristics or of rent and household characteristics. The estimated equations represent a simplified and rather ad hoc specification of a complex decision-making process and a set of price and quantity interactions. They implicitly assume "weak separability" of the utility function. Thus, similar to Barnett's model, separability implies that the consumer chooses how to allocate his budget in two stages—first, how to allocate spending among housing and other goods, and second, given a level of rental expenditures, how to allocate this among the attributes of the housing bundle. For "lumpy" consumption decisions like housing expenditures, this assumption may be reasonably realistic.

Several specification problems remain, however. First, the equations contain no price terms. Because of this, the coefficients of the included variables may be biased. If the price of an attribute is not constant across the sample geographic area, households presumably respond to these price variations. Given the present sample size, however, it was not feasible to assess potential price variation over many, relatively small geographic areas. (Subsample equations for "ghetto" and racially mixed submarkets provide estimates of an "average" price in only two geographic submarkets.)

A second issue is the simultaneity present in the consumption of housing attributes. Household decisions concerning amounts of housing attributes purchased face a common budget constraint. Decisions are likely to involve

TABLE 9.5 Component Demand Functions: Coefficient of Race of Household Variable in Pittsburgh

Dependent Variables	*XHBLACK (Black Household)*		
	Coefficient	*t-statistic*	R^{-2}
Persons per room	.016	1.31	.65
Dishwasher and/or disposal	−.011	0.77	.02
Central heat	−.036	1.20	.05
Inferior or no heat	.149**	6.05	.08
Adequate light and ventilation	−.107**	3.63	.04
Adequate kitchen facilities	−.021*	1.96	.03
Adequate plumbing	−.067**	2.92	.05
Building age	1.96*	2.27	.04
Average surface and structural quality	−.14**	6.73	.12
Total rooms (1n)	−.018	1.13	.28
Square feet per room	1.450	0.97	.04
Large multifamily structure	−.103**	4.89	.09
Single-family detached house	.048*	2.09	.07
Quality of blockface landscaping	−.267**	4.70	.04
High-quality blockface	−.205**	7.20	.06
Many high-quality features	−.024*	2.03	.01
Overall evaluation rating	−.267**	7.22	.14
Overall exterior cleanliness	−.339**	7.15	.11
High-quality kitchen	−.030†	1.80	.03
Number of components failed (1n)	.192**	5.65	.08
Basic plumbing and kitchen facilities	−.089**	3.10	.04
Median age of housing stock	.439	0.75	.00
Proportion Standard Dwelling Units	−.033**	6.47	.04
Air-conditioning	−.035†	1.87	.04
Quality of neighborhood landscaping	−.182**	9.59	.09
Problems with litter and trash	.165**	14.74	.18
Problems with crime	.117**	11.91	.10
Quality of recreation facilities	−.184**	10.45	.11
Poor quality elementary schools	.045**	14.97	.15
Neighborhood quality index	−7.94**	18.00	.22
Neighborhood problems index	5.36**	15.08	.16
Poor quality neighborhood services	4.57**	15.78	.19
Central city	.189**	6.18	.03

SOURCE: Merrill (1976).
NOTE: Independent variables include, in addition to XHBLACK, the natural log of income, level of education, age, family status, employment status, and household composition.
**t-statistic significant at 0.01 level.
*t-statistic significant at 0.05 level.
†t-statistic significant at 0.10 level.

tradeoffs among attributes. The nature of these tradeoffs is likely to differ as a function of household characteristics. Also, to the extent that groups of attributes are jointly produced, more complexity is introduced.

Despite these limitations, the results strongly suggest marked differences in consumption patterns for minority households and households living in primarily minority submarkets. For example, Table 9.5 shows the estimated coefficients for black households in Pittsburgh (income and the demographic variables mentioned above were also included but are not shown). Black households in Pittsburgh consume lesser amounts of numerous dwelling unit attributes than similar white households. Blacks generally consume less dwelling unit "quality" and less neighborhood "quality." However, blacks consume as much space as white households and have the same average number of persons per room. (These results are quite similar to those found by Kain and Quigley, 1975, for St. Louis.)

Black households are less likely to live in dwelling units that have basic facilities. For example, blacks are less likely to have both adequate plumbing and adequate kitchen facilities, are less likely to have adequate light and ventilation, and are more likely to have inferior heating facilities. Black households fail, on average, 1.2 more of the 15 components of minimum standard housing than white households; this represents 0.7 of the sample standard deviation of the number of components failed. The average rating for surface and structural quality for blacks is lower than that for whites; similarly, the overall evaluation rating and the rating for general exterior cleanliness are lower for black households.

The most striking differences between black and white households are seen for neighborhood attributes. Blacks are 20% less likely to have a high-quality blockface. The average rating for blockface landscaping is much lower than for white households. The average proportion of standard dwelling units in the census tract is 3.3% lower for blacks; this difference is over one-third of the standard deviation for this variable. Average black consumption of neighborhood quality, including landscaping, recreation facilities, elementary schools, absence of crime problems, and absence of litter and trash problems, are all considerably lower than for white households.

Finally, black households are about 19% less likely to live in the suburbs than similar white households. A comparison of the mean values of housing attributes in the central city and suburbs shows that both dwelling unit and neighborhood quality are lower in the central city. Clearly, differences in locational choice (or opportunity) introduce structural differences between the housing stock available to black and white households.

The results presented in Table 9.5 are illustrative only. Two more sets of equations were estimated for Pittsburgh: first, including expenditures rather than income, and second, replacing household dummies with those for submarkets. It is interesting to note that when expenditures are controlled, blacks still consume much less quality but purchase more space. Finally, a parallel analysis was carried out for Phoenix. In general, the results for unit quality and neighborhood attributes are similar: Hispanic households consume less than comparable whites. Two contrasts with Pittsburgh are evident, however: Hispanics purchased slightly less space than whites and were equally likely to live in the suburbs.

NOTES

1. In addition, separate hedonic indices were estimated in order to compare housing allowances in Pittsburgh and Phoenix with public housing and Section 23 and Section 236 housing (see Mayo et al., 1980a).

2. For simplicity, the tenure variables are assumed to be orthogonal to X and Z.

3. Comparison of R^2 for different forms of the dependent variable is not legitimate. Comparison of log likelihoods suggests that the semilog form is more powerful (see Merrill, 1980, p. 29).

10

Earmarked Income Supplements

JOHN E. MULFORD

☐ THOSE WHO ENROLLED in the Housing Assistance Supply Experiment's allowance program were offered monthly cash payments that began as soon as their eligibility was certified and their dwellings met the program's housing standards, and that continued for as long as they passed regular eligibility and housing recertifications. In the two HASE sites combined, more than 20,000 households became allowance recipients during the first five program years. Drawing on their participation records and on the HASE countywide surveys of households, this chapter describes how the program affected those who were recipients at the end of the third program year, a group that we think fairly represents the characteristic mix of participants in a mature, permanent program.[1]

Our assessment of program effects focuses on changes in housing consumption and changes in household budgets. That focus reflects the program's dual purpose. It was designed both to increase the housing consumption of low-income households that occupied substandard housing and to increase nonhousing consumption (or, equivalently, ease housing expense burdens) of those already occupying standard housing. The housing allowance offices' housing standards combined with recipients' housing preferences to determine the balance that was achieved between the two goals.

The chapter begins with estimates of the program's effect on recipients' housing consumption and housing quality and a discussion of how those increases occurred—through repairs and moves. It then examines how recipients allocated their allowances between housing and other consumption, focusing on the causes—allowances, housing standards, and other program features—and on the resulting housing expenditure burdens. Finally, the chapter compares housing allowances with public housing and unrestricted cash grants in terms of consumption increases (housing and other) achieved per program dollar expended.

Mulford et al. (1982) give a full account of the analytical methods and estimation details underlying the results presented here. The analysis used program records for 1,848 renter and 763 homeowner recipients in Brown County and 1,945 renter and 2,056 homeowner recipients in St. Joseph County.

HOUSING CONSUMPTION CHANGES

We estimate that the allowance program caused recipients to consume about 8% more housing than they would have consumed without the program and that the proportion living in dwellings of standard quality increased from about half to over four-fifths.[2] Renters (who had a program effect of 8.0% with a 2.6% standard error) achieved their housing improvements partly by repairing and partly by moving; owners (8.4% program effect with a 2.0% standard error) achieved theirs almost entirely by repairing their homes.

The modest program-induced increase in housing consumption evaluated in terms of market rents and repair expenditures contrasts sharply with the 30-point increase in the percentage of recipients occupying dwellings that would pass the HAO's housing standards. The large increase in standard housing for recipients is possible without a large increase in their housing consumption because many violations of the housing standards are inexpensive to remedy. With housing allowances as an incentive, enrollees fixed many such defects that they would have otherwise ignored. Not all recipients lived in standard housing, even though they had to occupy standard housing to begin receiving payments, because defects appeared in their dwellings between annual inspections.

The allowance program affected recipients' housing consumption by altering both their repair and moving behavior. Moving offers much wider possibilities for housing change than does repairing. When a household moves, it can change all of the attributes of its housing—space, quality, and location. Repairs, on the other hand, address primarily dwelling quality. Adding rooms is expensive, and a repair cannot change the location of a dwelling.

Because their characteristics and circumstances differ greatly, owners and renters use much different combinations of moving and repairing to change their housing consumption. Because owners control their own repair policies, and because moving entails selling one house and buying another, they make the modest housing adjustments stimulated by the allowance program by changing their repair behavior rather than by moving. Renters have less influence over repair policies, and moves are less expensive for them, so they may make even modest housing adjustments by moving rather than improving a landlord's property at their own expense.

Enrollees' housing units were evaluated according to a 38-item checklist, which was derived from the current housing codes in the two sites, the Building Officials and Code Administrators model code, and minimum housing standards developed by organizations such as the American Public Health Association. The most common reasons for failing the evaluation were inte-

TABLE 10.1 Cost of Repairs Made by Allowance Recipients in Response to Evaluation Failures, by Tenure and Site

Site	*Percentage Repairing During a Year*	*Average Cost ($/year)* *Per Recipient Making Repairs*	*All Recipients*
	Renters		
Brown County	24	111	27
St. Joseph County	40	95	38
Average	32	103	33
	Owners		
Brown County	24	95	23
St. Joseph County	35	77	27
Average	29	86	25

SOURCE: Mulford et al. (1982: Table 2.4).

NOTE: Data include both repairs made by enrollees seeking to qualify for payments and those made by recipients in response to subsequent annual evaluation failures. Repair costs include unpaid labor evaluated at the minimum wage as well as cash expenditures. For renters, both landlord and tenant expenses and labor are included.

rior stairway hazards, lead-based paint hazards, and unsafe (i.e., broken or damaged) windows. Missing or inoperable toilets, washbasins, and bathing facilities and unsafe plumbing, heating, and electrical systems were also common. Various other hazardous conditions, inadequate kitchen facilities, and inadequate living space comprised the remaining defects found in recipients' enrollment dwellings.

Violations for HAO housing standards are often so easy and inexpensive to remedy that virtually all owner recipients and three-quarters of renter recipients whose dwellings failed the evaluation repaired the defects rather than moved. We call such repairs "required" because they remedy violations of the housing standards that were explicitly cited by the HAO's evaluators. About a third of the households receiving payments in the mature program did some required repairs during the course of a year, either in connection with their initial qualification for payments or in order to avoid suspension (see Table 10.1). Those required repairs cost nearly $100 on the average, which comes to about $30 per recipient when averaged over both repairers and nonrepairers.

Both owners and renters did other (nonrequired) repairs while receiving payments, but only owners did more than they would absent the allowance program. We call those repairs "voluntary" because they were not prompted by a housing evaluation—they occurred during the year between evaluations—and they did not affect allowance payment status during the period in which they occurred. Almost three-quarters of owners did some voluntary repairs each year. Averaged over all owner recipients, the annual cash expense was $403 per recipient (see Table 10.2). These figures include both repairs they would have made even without the program ($263 average for

TABLE 10.2 Cost of All Program-Induced Repairs Made by Owner Recipients, by Site

	Average Annual Repair Expense ($)				
	Total	With Program			Program-
Site	Without Program[a]	Required Repairs[b]	Voluntary Repairs[c]	Total	Induced Repairs[d]
Brown County	236	23	391	414	178
St. Joseph County	290	27	416	443	153
Average	263	25	403	428	165

SOURCE: Mulford et al. (1982: Table 2.6).

NOTE: Except as indicated, repair costs reported in this table do not include any allowance for unpaid labor. When valued at the minimum wage, such labor adds about 12% to repair costs in Brown County, 7% in St. Joseph County.

a. Estimated without-program repair expenses of year 3 owner recipients.

b. From Table 10.1; includes a small amount of unpaid labor, valued at the minimum wage.

c. Voluntary repairs equals total minus required repairs.

d. Total with-program minus total without-program repairs.

the two sites) and the voluntary repairs that were caused by the program ($140 average), but not the required repairs ($25 average). Comparing owners' total annual repair expenses while in the program to our estimate of their expenses without the program, we conclude that the program caused them to increase their cash outlays for repairs and improvements by $165 annually, partly required but mostly voluntary.

In addition to altering repair behavior, the allowance program affected both the timing of recipients' moves and the amount of their housing changes when they moved. It caused some households to cancel or delay moves that would have decreased their housing consumption, and it caused others to speed up moves that increased their housing consumption. When recipients moved, the program caused them to increase their housing consumption by more than they would have done in its absence.

The 40% of renter recipients at the end of program year 3 who moved after enrolling accounted for most of renters' increased housing consumption due to the program. The typical year 3 recipient who moved after enrolling increased his gross rent expenditure (in constant dollars) by 16.5%, whereas those who did not move increased their expenditures by only 1.1%, presumably because of repairs.

Even though moves offer more opportunity for housing change than do repairs, renter recipients who moved did not much change the attributes unique to moving—total space and location. Movers' destination dwellings had the same number of rooms (on average) as their origin dwellings, and the small fraction of movers who changed neighborhoods—a third in Brown County and a fifth in St. Joseph County—chose destination neighborhoods that resembled origin neighborhoods in terms of quality of buildings and landscaping, general cleanliness, and access to employment. Instead, movers' changes concentrated on habitable space (rooms meeting HAO standards) and dwelling quality (high ratings on HAO checklist items).

The program affected owners' moves much less than it did renters' moves. Few homeowners increased their housing consumption by making unscheduled (program-induced) moves to better houses, because transaction costs for owners are high. The program might have caused some to delay moving to less expensive housing (e.g., from single-family homes to apartments); but because homeowners move infrequently (fewer than 10 percent of all homeowners in our sites normally move each year), there were few moves to be delayed. Controlling for life cycle stage, homeowner recipients in program year 3 moved slightly less frequently than homeowners surveyed at baseline (preprogram), suggesting that the program delayed moves for perhaps 2% of the year 3 homeowner recipients.

HOUSEHOLD BUDGET ALLOCATION

For both renters and homeowners in our sites, housing consumption had a high budgetary priority; but once a minimum level of consumption was achieved, additional housing had a low priority relative to other forms of consumption. In the jargon of economics, that behavior corresponds to low income elasticity of demand for housing. We have estimated income elasticities of approximately 0.2 for renters and 0.5 for owners in our sites (see Mulford, 1979). With those elasticities, a 10% increase in income would cause only a 2% increase in renters' and a 5% increase in owners' housing consumption.

Without the aid of a housing allowance, the average renter recipient would have spent 49% of his gross income on housing; the average owner recipient would have spent 43% of his gross income.[3] By spending such large fractions of their incomes for housing, recipients occupied adequate or nearly adequate housing (i.e., housing that would meet or almost meet HAO standards) when they enrolled; therefore, the HAOs' standards did not force them to increase their housing consumption very much. And because their income elasticities of housing demand were so low, they did not voluntarily use much of the allowance payment for additional housing consumption.

Out of a $1,014 average annual allowance, renter recipients spent only $161 (16%) on additional housing; the remaining $853 (84%) offset housing expenses they would have incurred without the program, freeing an equal amount for increased nonhousing consumption. Homeowners divided their $781 average annual allowance into $165 (21%) for additional housing consumption and $616 (79%) for nonhousing consumption.

Had the allowance been an unrestricted cash grant, recipients would have used even less of the allowance for increased housing.[4] Table 10.3 shows that factors other than the allowance payment accounted for nearly half of the housing increase for renter recipients and seven-eighths of the increase for owners.

We expected housing standards, which are the major element differentiating housing allowances from unrestricted cash transfers, to cause most of the housing increase not caused by the allowance payments. That may be true for renters; for owners, other factors played a major role. We judge that housing standards caused virtually all of the required repairs done by recipients.[5] But required repairs, which amounted to $33 for renters and $25 for owners annually (see Table 10.1), account for only a small fraction of the housing increases caused by nonallowance factors.

However, the housing standards can also work through moves and voluntary repairs. Renter recipients who moved after failing a housing evaluation had to overcome more extensive and serious defects than those who repaired. Movers from failed dwellings averaged 3.6 defects compared to 2.4 for those who repaired. About 43% of movers from failed dwellings had failed the occupancy standard, compared with 15% of those who repaired. Remedying an occupancy failure usually requires at least one additional room, which adds about 9% to rent.[6] Those who were able to correct an occupancy failure without moving usually had adequate space but needed to improve its heating, lighting, ventilation, or privacy.

Some of the program-induced voluntary repairs done by owners may have been prompted by the housing standards. Assuming that no voluntary repairs were standards-induced gives a lower-bound estimate, which indicates that

TABLE 10.3 Increased Housing Consumption by Allowance Recipients, by Cause of Increase, Tenure, and Site

	Year 3 Average Amount ($)			*Percentage Distribution*		
		Attributed Cause			*Attributed Cause*	
Site	*Housing Consumption Increase*	*Allowance Payments*	*Other*[a]	*Housing Consumption Increase*	*Allowance Payments*	*Other*[a]
		Renters				
Brown County	159	88	71	100	55	45
St. Joseph County	162	78	84	100	48	52
Average	161	83	78	100	52	48
		Owners				
Brown County	178	17	161	100	10	90
St. Joseph County	153	26	127	100	17	83
Average	165	22	143	100	13	87

SOURCE: Mulford et al. (1982: Table 3.3).

a. Possible causes include the HAOs' housing standards, "Hawthorne effect" of participation in an experiment, housing inspections that call attention to incipient problems, and the moral pressure of receiving a *housing* allowance.

almost three-fourths of program-induced repairs are attributable to causes other than allowance payments and housing standards. But recipients probably do some voluntary repairs to prevent dwelling failures in the future. Comparing deficiency and voluntary repairs by type, McDowell (1979: 46) estimates that up to 15% of voluntary repairs may fix items that would have failed at the next annual evaluation.[7] Under that upper-bound assumption, housing standards caused half of the housing consumption increase for owners.

Even the upper bound leaves a third of owners' housing consumption increase unaccounted for. The HAO's data collection activities and advertising that stressed the housing objectives of the program may explain the residual increase. Regular housing evaluations and questions about repairs may have stimulated recipients to do more repairs so they would "look good" at the next evaluation. Calling the program the *Housing* Allowance Program and advertising it heavily as a means to help people with their housing might have increased repair expenditures in the following way. Many recipients—particularly elderly homeowners—would never have joined a welfare program, but they joined the housing allowance program because its advertising convinced them that the money was for a socially acceptable purpose—improved housing. After joining the program, they may have felt morally obligated to spend their allowances on housing even though they already met the program's housing standards.

Although allowances substantially increased recipients' incomes—by an average of 17% for homeowners and 25% for renters—those augmented incomes are still far below the average nonrecipient's income. Dividing recipients' housing expenditures by their allowance-augmented incomes yields housing expenditure burdens (see Table 10.4) that are well above a typical legislative standard of 25% (the maximum share of income a household should have to spend to acquire decent, safe, and sanitary housing). To achieve an average 25% burden would require quadrupling the current program's average allowance payment.

Most federal housing assistance programs actually calculate expense burdens differently, treating the federal subsidy as an offset to housing expenses rather than as an increment to income. As shown in the last column of Table 10.4, by that calculation, housing expense burdens for allowance recipients decrease to 28% for renters and 29% for owners. This measure has enjoyed wide use, probably because it was the natural one for public housing, the nation's oldest low-income housing assistance program. In public housing, the government supplies program participants with a housing unit and charges them a below-market rent equal to 25% of their income. Thus their with-program housing expense burden is 25%. For allowance recipients, subtracting the allowance payment from with-program housing expenditures and dividing by nonallowance income results in burdens slightly higher than 25%, because recipients get allowances equal to the standard cost of ade-

TABLE 10.4 Effect of Allowances on Recipients' Housing Expenditure Burdens, by Tenure and Site

	Housing Expenditure/Gross Income (%)		
		With Program	
Site	*Without Program*[a]	*Allowance Added to Income*[b]	*Allowance Subtracted from Housing Expense*[c]
		Renters	
Brown County	45	40	27
St. Joseph County	54	45	29
Average	49	43	28
		Owners	
Brown County	39	37	27
St. Joseph County	46	42	32
Average	43	39	29

SOURCE: Mulford et al (1982: Table 3.6).

NOTE: Entries are ratios of average housing expenditure to average gross income, expressed as percentages.

a. Estimated without-program housing expenditures of year 3 recipients, divided by their nonallowance gross incomes.

b. Actual year 3 housing expenditures, divided by gross incomes including allowances.

c. Actual year 3 housing expenditures minus allowances, divided by nonallowance gross incomes.

quate housing (R*) minus 25% of income, but many choose to live in dwellings that cost more than R*.

PROGRAM EVALUATION

As shown above, those who participated in the experimental housing allowance program benefited both with respect to housing consumption and budgetary relief. Their overall housing consumption increased by about 8%, nearly a third shifted from substandard to standard dwellings; and the funds available to recipients for nonhousing consumption increased by 41% for renters and 23% for homeowners. These are measures of program effectiveness, but they must be qualified by the fact that benefits were received by less than half of all eligibles and 80% of all enrollees. The "efficiency" for the program can be measured by comparing benefits actually bestowed with the program's cost. We have excellent data on program costs, and the benefits of housing allowances to participants are readily measured. We are confident that the incidental costs and benefits to nonparticipants are small, but we estimated them only indirectly.

Because housing allowances are of interest as an alternative to other policy instruments, we present parallel assessments of the efficiency of the public housing program, whose low-income participants live in dwellings

TABLE 10.5 Program Costs and Participant Benefits for Alternative Assistance Programs

Item	*Monthly Costs and Benefits Per Standard Case ($)*		
	Public Housing	*Housing Allowances*	*Unrestricted Grants*
	Program Cost to Deliver Equal Participant Benefit		
Benefit to participant	73.55	73.55	73.55
Administration and other	141.60	13.55	9.22
Total	215.15	87.10	82.77
	End Use of Benefit by Standard Participant		
Housing consumption	11.51[a]	11.51[a]	5.93
Other consumption	62.04	62.04	67.62
Total	73.55	73.55	73.55

SOURCE: Mulford et al. (1982: Table 5.1).

NOTE: The standard case is a renter recipient whose adjusted gross income is $4000 annually. Without the program, he would spend $144.16 monthly for housing. The public housing authority provides him with a dwelling whose market rental value is $155.67, the amount he would choose to spend if given a housing allowance. With an unrestricted grant, he chooses to spend $150.09. In fact, the typical public housing tenant has a lower income and is provided with a dwelling whose market rental value is $145.00.

a. Participants in the public housing and housing allowance program would evaluate this portion of the benefit at less than $11.51 because of constraints on its use.

owned and operated by local housing authorities but federally subsidized; and of a hypothetical program of unrestricted cash grants to low-income households, such as has been studied by the Office of Economic Opportunity and the Department of Health, Education and Welfare.

The public housing program fully specifies its participants' housing consumption; the housing allowance program merely sets minimum standards for housing consumption; and the unrestricted grant program leaves consumption choices entirely to recipients. These are the essential differences between the three policy instruments, although in each case other differences can be created by varying eligibility or entitlement standards.

To understand how program structure affects efficiency, we must apply all three programs to a standard case even though they may actually serve somewhat different populations and have somewhat different benefits schedules. We have designed a standard case such that each program delivers approximately the same amount of benefit ($73.50 per month) to typical renter recipient with an income of $4,000 (see Table 10.5). We used data from HASE and from a concurrent study of public housing (Mayo et al., 1980a, 1980b) to estimate (a) the total program cost entailed in supplying the benefit, (b) how the recipient divides it between housing and other consumption, and (c) how other households not in the program are affected. Implicitly, we assumed that all three programs were operated in the HASE sites and that they were open only to the renters eligible for assistance under HAO rules.[8]

The upper panel of Table 10.5 reveals the main consequence of the differences between the programs: To deliver the specified benefit, the public housing program incurs 2.5 times the cost incurred by the housing allowance or unrestricted grant program. The main reason, according to Mayo et al. (1980b), is that public housing authorities are inefficient real estate developers; they pay about $2 for every $1 of housing service they produce. The housing allowance program spends about $14 per recipient month to administer means tests and housing evaluations, whereas the unrestricted transfer program would need only the means test, estimated at $9 per recipient month.

The lower panel of the table shows how the recipient would use his benefit. In order to make the programs comparable, we assumed that the public housing program provided somewhat better dwellings than it actually does, so that both the public housing and housing allowance programs caused recipients to increase their housing consumption by about $12, taking the rest of their benefit in cash. The unrestricted grant recipient would choose to spend about $6 extra on housing, using the rest for other purposes.

The costs and benefits of these programs are not necessarily limited to those participating in them; they may affect the housing choices available to others in the same marketplace. Below we compare the marketwide consumption changes caused by the three programs, dividing consumption into housing and other goods because we are evaluating the housing allowance program.

TABLE 10.6 Consumption Changes Caused by Alternative Assistance Programs, by Participation Status and Type of Consumption

	Consumption Change per Assistance Program Dollar ($)		
Type of Consumption	*Public Housing*	*Housing Allowances*	*Unrestricted Grants*
		Participants	
Housing	.05	.13	.07
Other	.29	.71	.82
Total	.34	.84	.89
		Nonparticipants	
Housing	.03	.01	(a)
Other	.03	−.02	−.01
Total	.06	−.01	−.01
		All Households	
Housing	.08	.14	.07
Other	.32	.69	.81
Total	.40	.83	.88

SOURCE: Rydell and Mulford (1982: Table 4.1).
NOTE: Population characteristics for all programs are averages across HASE sites.
a. Rounds to zero; calculated value is −.002.

Table 10.6 shows how both housing and other consumption are changed by each program and includes separate accounts for participants and nonparticipants. Participants' consumption increases per program dollar are always less than 1.0 because of administrative and other nonsubsidy program costs. Delivering unrestricted cash grants entails the least administrative cost; therefore, they provide the greatest consumption increase per program dollar—0.89 for participants. Housing allowances, which require modest housing standards enforcement costs, deliver almost as much subsidy to participants as do unrestricted cash grants. Public housing's high development costs combine with administrative expenses to absorb nearly two-thirds of the federal subsidy without benefit to participants; only 0.34% of each program dollar went for participants' consumption increases.

The high development costs of public housing have been well known for years (see Rydell and Mulford, 1982: 1-2). But supporters of public housing have argued that the addition of the public housing units benefited nonparticipants as well as participants. If an increased supply of housing leads to lower marketwide prices, nonparticipants' benefits could outweigh the higher cost per participant of the supply-side strategy. In contrast, housing allowances and unrestricted cash grants stimulate demand, possibly pushing up prices for nonparticipants.

Modeling the market effects of each program using HASE and other data, Rydell and Mulford (1982) conclude that none of the programs has much effect on nonparticipants' consumption of housing or other goods. In response to supply programs, the market response offsets most of the new housing units; either new construction is deferred or demolitions increase. In response to demand programs, vacancy rates fall, but housing prices increase only slightly, causing only a small reduction in nonparticipant demand. Nonparticipant housing consumption actually increases with demand programs, because future participants increase their consumption in anticipation of joining the program and, with housing allowances, housing standards cause some recipients to consume above-normal housing after they terminate from the program.

Altogether, housing allowances cause about twice as much increase in marketwide housing consumption per program dollar as either of the other programs. As compared to public housing, housing allowances cause both more housing consumption increase and more nonhousing consumption increase per program dollar. As compared with unrestricted cash grants, housing allowances cause more housing consumption increase but less increase in the consumption of other goods, and less total consumption increase. Because increased housing consumption (particularly if it rids dwellings of health and safety hazards) has social as well as individual value, the extra housing consumption caused by housing allowances might outweigh the cost of housing standards enforcement that distinguishes housing allowances from unrestricted cash grants.

NOTES

1. The program reached steady-state size by the third program year. In subsequent years, the mix of household types changed slowly in favor of the elderly who tended to remain eligible and in the program longer than younger households.

2. Lacking a formal experimental control group, we estimated what recipients would have done without the allowance program (in terms of their rents and repair expenditures) by using contemporaneous survey data on nonrecipients.

3. These are the fractions of income recipients spent on housing just prior to enrollment.

4. We calculate the allowance payment's effect on recipients' housing consumption as the ratio of allowance-augmented income to nonallowance income, raised to the power of the income elasticity.

5. Although some of those repairs might have been done eventually without the program, they tended to remedy defects, such as broken electrical switch plates, stuck windows, and chips of paint on the ground, about which recipients and their landlords were either unaware or unconcerned.

6. We computed the percentage increase in rent caused by adding a fifth room (occupancy-deficient dwellings averaged four rooms) from hedonic indexes for Brown County (Barnett, 1979) and St. Joseph County (Noland, 1980) and averaged them.

7. Even if those 15% of voluntary repairs had been required at year end, had they not been done voluntarily during the year, attributing them to housing standards probably misplaces causality for some of them—that is, recipients would have done some of them without standards.

8. The models and assumptions underlying the comparisons of alternative programs are detailed in Rydell and Mulford (1982).

11

Price Effects of Housing Allowances

C. PETER RYDELL and C. LANCE BARNETT

☐ THE PRIMARY PURPOSE of the Housing Assistance Supply Experiment (HASE) was to learn how a full-scale housing allowance program would affect local housing markets; in particular, the price of housing services. Knowing the size of program-induced price increases is important because large increases would both disrupt the housing market and divert program subsidies from the intended recipients. This chapter presents the experiment's findings, which show that the price effects were negligibly small.

In the housing allowance program households receive the difference between the standard cost of adequate housing (which varies with household size) and one-fourth of the household's income, provided they choose housing that meets the program minimum requirements. The "standard cost" used in the allowance payment formula is an estimate of the full market rent of rental dwellings that meet program standards. It does not depend on the rent of the specific dwelling chosen by the allowance recipient. In fact, recipients can move from one standard dwelling to another without losing benefits. Recipients, like other households, therefore have an incentive to seek bargains. If they can find a dwelling meeting program requirements and renting for less than the standard cost, they keep all the rent savings. Consequently, the program motivates tenants to pay no more than the market price for their housing services. However, since the supply of housing services is less than perfectly elastic, at least in the short run, program-induced demand will clearly cause the market price to increase. The question is, how much?

AUTHORS' NOTE: *This chapter benefited greatly from analyses by Ira S. Lowry, John E. Mulford, and Kevin Neels. For more detail on price effects, see Rydell, Neels, and Barnett (1982).*

PRICE INCREASE FEARS

Before the experiment many people thought housing allowances would cause serious price increases, arguing that the market would not be able to accommodate the increased demand for housing. For example, commenting on the probable impact of a housing allowance program, DeLeeuw and Ekanem (1971) concluded that "subsidizing the demand for low-income housing would drive up rents" (p. 817). In testimony to the U.S. Congress (1972), Henry Aaron summed up the judgment of many housing experts and predicted that a housing allowance program would cause a 10% price increase.

The National Bureau of Economic Research's housing market simulation model predicted that a full-scale housing allowance program operated in either Pittsburgh or Chicago would have serious price effects. One-fifth of the housing market would experience price increases exceeding 10%, and one-tenth would experience price increases of more than 20% (Kain and Apgar, 1977: Table 9.5).

The initial application of the Urban Institute's (UI) housing market model concluded:

> In seven of the simulations . . . housing prices for recipients of the housing allowance rise. They rise by more than 10 percent in five of the eight cases. . . . The results thus do confirm the fear that a large scale housing allowance program carries the danger of upward pressure on prices [DeLeeuw and Struyk, 1975: 131].

In subsequent applications, the UI model predicted that the housing allowance program actually being run in Brown County would cause the price of housing services for recipients to rise by 4%-9% and that the program actually being run in St. Joseph County would cause the price of housing services for recipients to rise by 20%-27% (Vanski and Ozanne, 1978: Tables 3, 5, and A.2).[1]

The fears of program-induced price increases were shared by nonacademics. In 1974 HUD initiated a housing assistance program (the Section 8 Existing Housing program) that closely resembles the one tested by HASE. However, rather than trust the market to set rents, HUD chose to set rents administratively using direct negotiations between local housing authorities and landlords. Further, HUD allowed the subsidies to vary with actual rents. As a consequence of those choices, the average price increase for dwellings occupied by recipients of that program's subsidies was 26% (see Rydell et al., 1980).

THE EXPERIMENTAL EVIDENCE

The Supply Experiment operated full-scale housing allowance programs in Brown County, Wisconsin (whose central city is Green Bay) and St. Joseph County, Indiana (whose central city is South Bend). Those locations were chosen because they had contrasting market conditions. Brown County had a tight housing market (4% rental vacancy rate), while St. Joseph County had a loose housing market (10% rental vacancy rate). In this section we examine empirical evidence for program-induced price increases during the first three years of program operation, the period when the program had its maximum effect on prices (see the theoretical evidence below). While we find no evidence of marketwide effects, we do find that submarket prices increased slightly because of the allowance program.

Four annual surveys addressed to households of a fixed set of dwellings in each location obtained information on the rents paid by occupants of a total sample of about 2300 dwellings. The surveys span roughly the first three years of program operations. We linked records for individual dwellings whose occupants responded in two or more annual surveys and calculated the annual percentage change in gross rent[2] for each pair of linked records. Because most dwellings change very little from year to year, the average percentage change in rent closely approximates the average percentage change in the price paid per unit of housing service.[3]

MARKETWIDE PRICE INCREASES

During the initial three years of the allowance program, the average price of rental housing services rose 26% in Brown County and 19% in St. Joseph County. However, those price increases were not caused by the allowance program but are attributable to background price inflation in the economy. During the initial three years of the Brown County program the Consumer Price Index for the north-central United States rose 27.4%, and during the initial three years of the St. Joseph County program it rose 22.9% (U.S. Department of Labor, 1972-1977). Moreover, during the same periods, the cost of producing housing services rose 27.2% in Brown County and 23.4% in St. Joseph County (Rydell et al., 1982). Fuel and utility costs rose most (due to the energy crisis), whereas real estate taxes rose least, but the cost of all inputs to the production of housing services rose by essentially the same amount as the Consumer Price Index.

Overall, the cost increases exceed the price increases (see Table 11.1), implying that the allowance program did not cause price increases. The observed relationship between cost and price increases is not unique to Brown and St. Joseph counties; national data covering the 1970s display the same pattern (see Lowry, 1982c).

SUBMARKET PRICE INCREASES

The absence of program-induced price increases marketwide does not rule out price increases in submarkets. The program's impact is focused on the areas where recipients choose to live. (Program regulations prevent them

TABLE 11.1 Marketwide Rent and Cost Changes During the First Three Years of the Housing Allowance Program

	Percentage Change Over Three Years				
	Gross Rent per Dwelling[a]	*Housing Services per Dwelling*[b]	*Rent per Unit of Housing Service*[c]	*Cost per Unit of Housing Service*[d]	*Difference Between Rent and Cost Change*[e]
Brown County	24.4	−1.6	26.0	27.2	−1.2 (±0.6)
St. Joseph County	18.3	−1.0	19.3	23.4	−4.1 (±0.9)

SOURCE: Estimated by HASE staff from linked records of annual surveys of rental dwellings and from price indexes of components of annual housing costs during 1974, 1975 and 1976 in Brown County and 1975, 1976 and 1977 in St. Joseph County. For additional details see Lindsay and Lowry (1981), Follain and Malpezzi (1980), and Rydell et al. (1982).

NOTE: Numbers in parentheses are standard errors of estimate due to sampling variability in measuring rent changes. They indicate a 66% confidence interval. Doubling the standard errors indicates a 95% confidence interval.

a. Gross rent consists of tenant payments to landlords (contract rent) plus any cost of fuel and utilities which tenants pay directly.

b. Housing services per dwelling decline due to normal deterioration.

c. Column 1 less Column 2.

d. Price indexes for components of rent weighted by the size of the component.

e. Column 3 less Column 4.

from living in substandard housing and low incomes prevent them from living in the very best housing.) If the submarket used by recipients is sufficiently insulated and small enough, the program could cause large price increases there.[4]

However, Table 11.2 shows that annual rent increases for dwellings currently occupied by program participants were only a few percentage points higher than for dwellings occupied by nonparticipants. Moreover, the difference cannot be attributed entirely to price increases, since at least part was due to an increase in the quantity of housing services resulting from program-induced repairs (about three-eighths of all renter recipients repaired their dwellings to qualify for allowance payments).

The evidence in Table 11.2 indicates that the extra demand for housing caused by the allowance program does not get focused very much. Independent evidence that allowance recipients shop for their housing throughout much of the housing market comes from the housing evaluations done by the housing allowance offices (HAO) in each county. The HAO inspects the dwellings potential recipients occupy or are considering occupying to determine whether they meet program standards. By mid-1979, 68% of the rental units in Brown County and 51% of the rental units in St. Joseph County had been evaluated at least once.

We cannot estimate precisely the extent to which the differential annual increases in Table 11.2 accumulate, because the dwellings occupied by program participants change from year to year. However, we judge that the dif-

TABLE 11.2 Rent Changes for Recipients' and Nonrecipients' Dwellings During the First Three Years of the Housing Allowance Program

	Average Annual Percentage Change in Gross Rent		
Period[a]	*Recipients' Dwellings*[b]	*Nonrecipients' Dwellings*[c]	*Differences*
		Brown County	
Period 1	8.8	5.6	3.2 (1.7)
Period 2	12.2	9.6	2.6 (1.3)
Period 3	9.2	7.2	2.0 (1.1)
All periods	9.9	7.4	2.5 (0.8)
		St. Joseph County	
Period 1	7.4	4.3	3.1 (2.5)
Period 2	9.5	7.4	2.1 (2.1)
Period 3	6.3	5.3	1.0 (1.5)
All periods	7.5	5.5	2.0 (0.9)

SOURCE: Estimated by HASE staff from linked records of annual surveys of rental dwellings. For additional details, see Lindsay and Lowry (1981).

NOTE: Numbers in parenthesis are standard errors of estimate due to sampling variability in measuring rent changes.

a. In Brown County, where program enrollment began in June 1974: Period 1 = December 1973-December 1974; Period 2 = January 1975-December 1975; Period 3 = January 1976-July 1977. In St. Joseph County, where program enrollment began in April 1975; Period 1 = November 1974-December 1975; Period 2 = January 1976-December 1976; Period 3 = January 1977-July 1978.

b. Average rent changes for dwellings occupied by allowance recipients during at least part of an observation period.

c. Average rent changes for dwellings not occupied by recipients during an observation period.

ferential rent increase is not cumulative. Rather, it occurs when or shortly after participants enter the program, as shown by the decline in the annual differential rent increases during the study period. (The later years have relatively fewer new participants and relatively more continuing participants.)

Table 11.3 provides direct evidence on the rent increases experienced as a dwelling's occupants join the allowance program. The table distinguishes between the dwellings not requiring repairs to meet program standards and those that were repaired. In both counties the rent increases for the dwellings requiring repairs exceeded that for those requiring no repairs, indicating that part of the average rent increase caused by the program is due to quantity rather than price increases.

The rent increases for dwellings requiring no repairs provide unbiased estimates of the immediate price impact of the housing allowance program: a 1.6% increase in Brown County and a 0.7% increase in St. Joseph County. By any standard such price increases are small. Moreover, those immediate increases are all the allowance program appears to cause. The evidence just reviewed strongly suggests that the housing allowance program does not cause annual increases in price that accumulate over time, but rather causes only a small one-time increase in price at the start of the program.

TABLE 11.3 Rent Changes for Dwellings When Their Occupants Join the Allowance Program

Repair Status	*Average Monthly Gross Rent ($)* Preprogram Rent	Program Rent	*Average Rent Increase (%)*
		Brown County	
No repair required	164	167	1.6
Repair required	151	155	2.5
All cases	159	162	1.9
		St. Joseph County	
No repair required	157	158	0.7
Repair required	152	155	1.7
All cases	155	156	1.2

SOURCE: Brown and St. Joseph County Housing Allowance Office records. For additional details, see Rydell et al. (1980).

NOTE: The entries in this table are for renter households who did not move when they entered the allowance program. The households reported their contract rents when they enrolled and again when their dwellings were certified for occupancy; the HAO estimated the cost of tenant-paid utilities from standard tables. The average interval between the enrollment interview and first certification was 1.6 months in Brown County and 2.1 months in St. Joseph County.

AN EXPLANATORY MODEL

The Supply Experiment also constructed models of how the program affected demand, how supply responded to the increased demand, and how prices changed as a consequence.[5] Our purpose here was not to predict precisely the program's effect on prices, but to ensure that our understanding of processes at work in a housing market were consistent with the observed outcomes. In fact, they were. The models predict virtually no effect on marketwide prices and a small effect on prices for housing demanded by program participants.

PROGRAM-INDUCED DEMAND

In modeling the demand shifts caused by the allowance program we first divide the rental housing market into two submarkets: the recipient submarket and the nonrecipient submarket. We define the recipient submarket as the set of dwellings that recipients can afford and that either meet program standards or could be brought up to standard inexpensively. We used HAO records to estimate the size of the recipient submarket. The HAO conducted evaluations not only for dwellings already occupied by program participants but for dwellings they considered possible residences. As mentioned earlier, during the first five program years, 68% of all rental dwellings in Brown County and 51% in St. Joseph County were evaluated at least once. The remainder of the rental inventory in each location constitutes the nonrecipient submarket.

We estimate that the program caused demand for rental housing service in the recipient submarket to increase by 4.6% in Brown County and 5.6% in St. Joseph County.[6] Most of the increase was due to the small proportion of recipients who shifted their housing demand from the nonrecipient to the recipient submarket because their enrollment dwellings were irremediably substandard. The much larger number of recipients who did not change submarkets contributed only their modest program-induced demand increases to the recipient submarket total.

Demand in the nonrecipient submarket decreased by an estimated 6.0% in Brown County and 3.5% in St. Joseph County as potential recipients living in irremediably substandard dwellings moved to the recipient submarket.[7] The marketwide demand increases were very small—about one-tenth of one percent in each site. (The marketwide change is the weighted sum of the submarket changes, which have opposing signs.) Although at its equilibrium level the allowance program served about 15% of all renters in each site, those in the program increased their housing consumption by only 8% on average—not enough to change marketwide totals substantially.

SUPPLY RESPONSE

When the aggregate demand for housing service in a market or submarket increases, the price per unit of that service rises enough to clear the market during the short run, while the housing inventory is fixed. However, landlords soon notice such price increases and find it profitable to expand the inventory. As the inventory expands, prices fall and consumption expands until the market reaches a new equilibrium.

We distinguish three ways in which the supply of rental housing services responds to demand shifts. First, some existing rental dwellings are repaired to meet allowance program standards. Second, the inventory of rental housing changes by new construction, demolition, or conversion, including shifting dwellings between rental and ownership markets. Third, the occupancy rate (fraction of housing occupied) within the rental inventory rises or falls, accommodating more or less consumption.

Because each of the three supply responses occurs at different rates, our model estimates the time path of each response separately. The total supply response then equals their sum. The results for the recipient submarket are plotted in Figures 11.1 and 11.2, along with the time path of the demand shift to which supply is responding.

The pattern is similar in the two counties. In both locations, the demand increase levels off in less than five years, as allowance program enrollment reaches its equilibrium level. The repair response grows as the program grows but is always relatively small. The inventory change increases steadily but slowly, each year accommodating more of the program-induced demand shift but taking many years to eliminate the difference between desired sup-

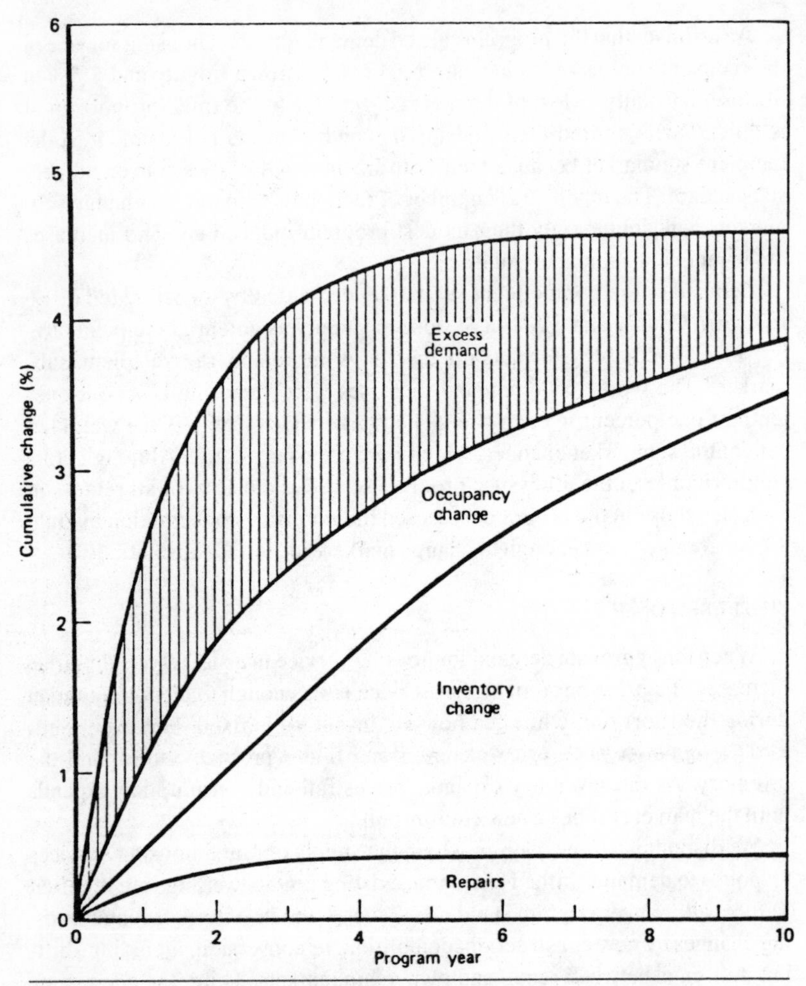

FIGURE 11.1 Demand Shift and Supply Response Caused by the Allowance Program: A Model of the Rental Housing Market, Recipient Submarket, Brown County

ply and actual supply. The occupancy rate adjustment accommodates between one-third and two-thirds of the demand change left after the repair and inventory responses.

The occupancy change is smaller in the tight Brown County market than in the loose St. Joseph County market because of the relatively high initial occupancy rate there (96%). Since occupancy rates can never, by definition, exceed 100%, the higher they are initially, the less increased demand they can absorb. By way of contrast, St. Joseph's County's initial occupancy rate was low (90%), so changes in it could absorb a greater proportion of the demand increase.

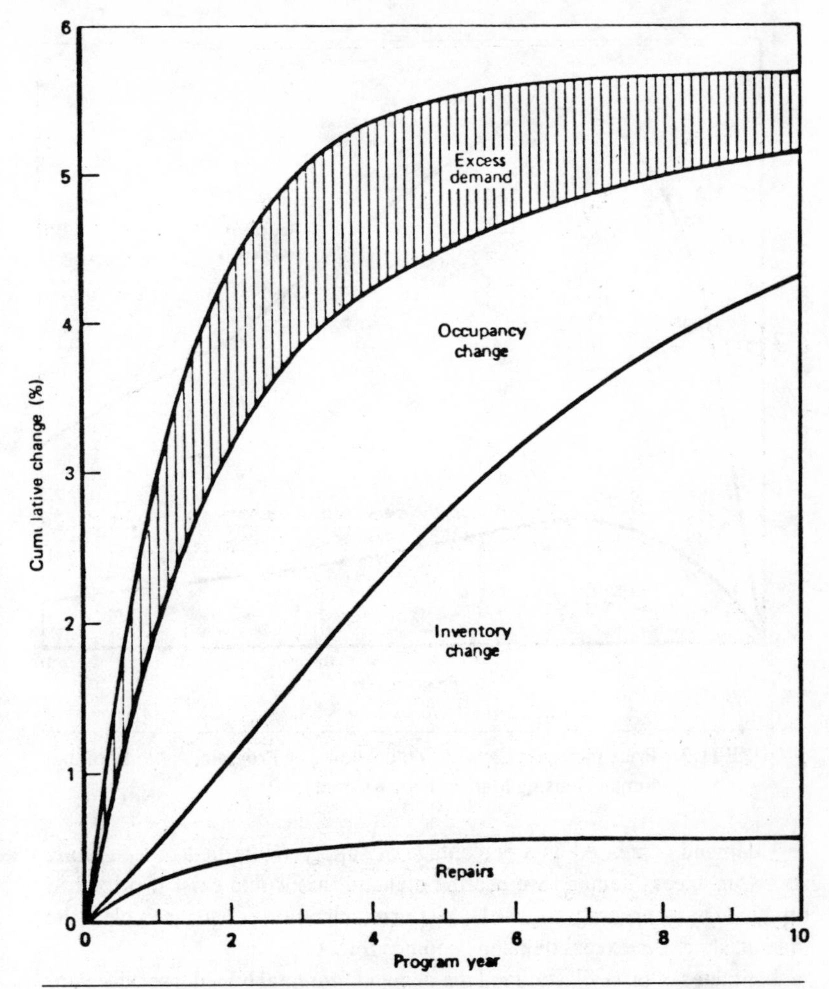

FIGURE 11.2 Demand Shift and Supply Response Caused by the Allowance Program: A Model of the Rental Housing Market, Recipient Submarket, St. Joseph County

PROGRAM-INDUCED PRICE CHANGES

The supply response in our model lags the demand changes. Consequently, to clear the market at any particular time requires price changes. The price of rental housing services must rise in the recipient submarket to eliminate short-run excess demand and must fall in the nonrecipient submarket to eliminate short-run excess supply.

With short-run excess demand, the percentage price increase needed to clear the market (by reducing the amount demanded) is the product of two numbers: the inverse of the price elasticity of demand and the short-run ex-

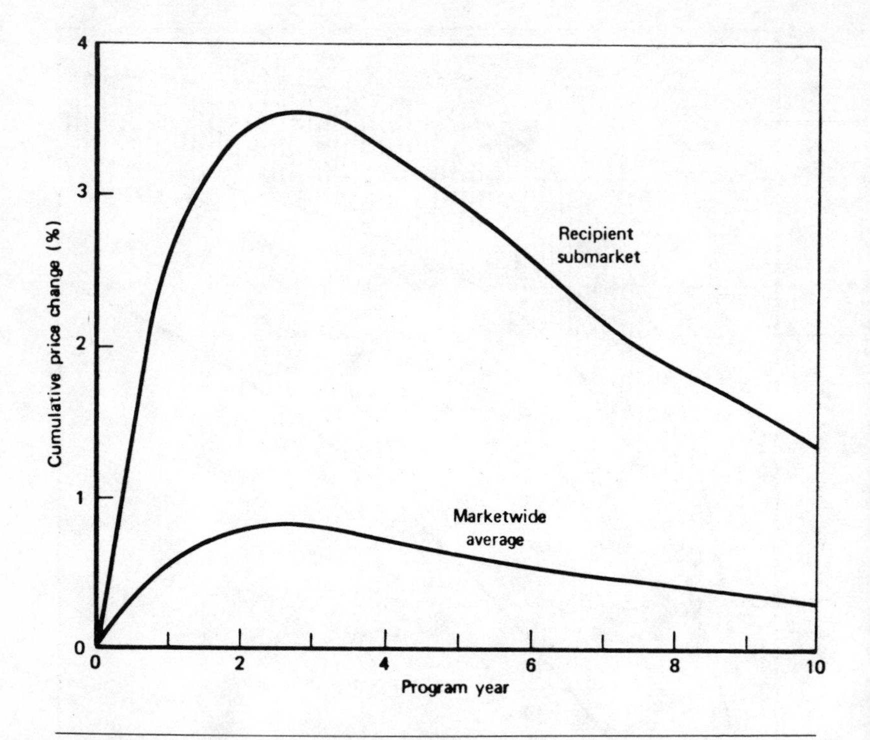

FIGURE 11.3 Price Increases Caused by the Allowance Program: A Model of the Rental Housing Market, Brown County

cess demand expressed as a percentage of supply. (Note that in estimating short-run excess demand we use the demand that would exist if price had remained at its preprogram level. Of course, once price changes to clear the market, short-run excess demand becomes zero.)

Estimates of price elasticity of the demand for rental housing service vary considerably, ranging from 0.17 to 1.28 in absolute value (Mayo, 1981). However, most of the estimates lie between 0.3 and 0.7, with a central tendency of 0.5. In our model, we use a value of 0.5, implying that a 10% increase in market price causes a 5% reduction in consumption.

To estimate the percentage price increase, we multiplied the inverse (2.0) of that elasticity by the percentage difference between supply and demand in each submarket of our two experimental sites (the short-run excess demand in Figures 11.1 and 11.2). For example, in the recipient submarket of Brown County during program year 1, we estimated that short-run excess demand equaled 1.3%. The price increase needed to clear the market was therefore 2.6%. The results for the first 10 program years in each site are plotted in Figures 11.3 and 11.4. In each county the maximum price increase occurs

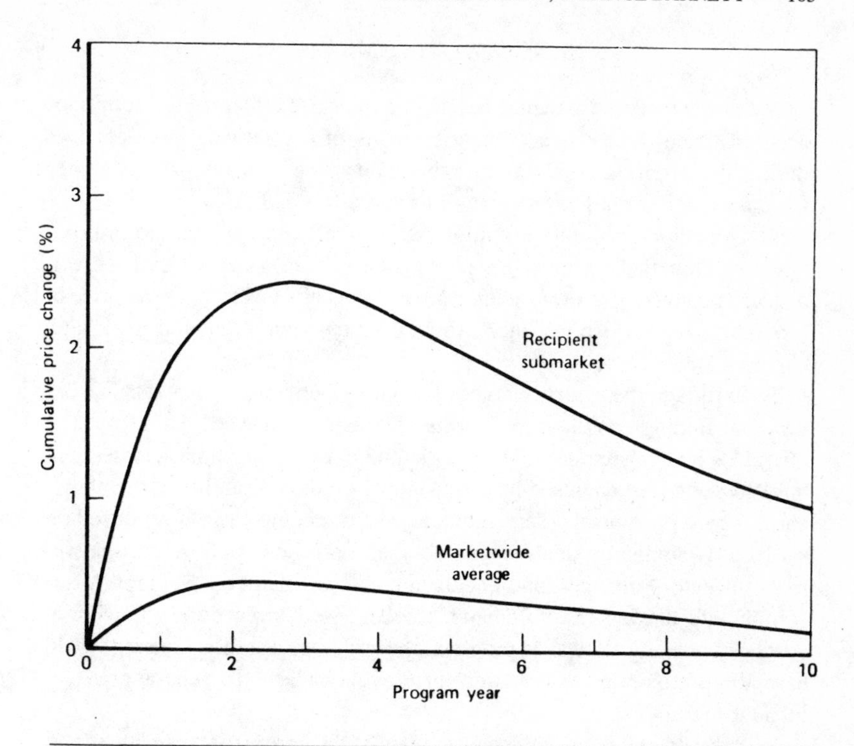

FIGURE 11.4 Price Increases Caused by the Allowance Program: A Model of the Rental Housing Market, St. Joseph County

between program years 2 and 3; thereafter, prices gradually return to their preprogram values as supply approaches demand.

However, the price changes are small. In the recipient submarkets, we estimate a maximum increase of 3.5% for Brown County and 2.5% for St. Joseph County. In the nonrecipient submarkets, we estimate maximum price decreases of 5.0% in Brown County and 2.4% in St. Joseph County. Marketwide, price changes never exceed 0.8% in Brown County and 0.5% in St. Joseph County.

The program-induced price changes predicted by our model exceed those identified by the empirical evidence presented earlier, primarily because our model does not allow nonrecipients living in the recipient submarket to move to the nonrecipient submarket and enjoy its lower prices there. If such movement occurred, modeled prices would not rise by as much in the recipient submarket and would not fall by as much in the nonrecipient submarket, which would strengthen our finding: The allowance program caused only small price changes.

CONCLUDING COMMENTS

We have presented evidence bearing on the price effects of the allowance program from three sources: measurements of marketwide price changes during the program, analyses of submarket price changes, and estimates from a model of market processes calibrated to conditions in each experimental location. Each source of information entails ambiguities and approximations. Nonetheless, only one conclusion is consistent with all the evidence: The housing allowance program had virtually no effect on the price of rental housing service in either Brown County's tight market or St. Joseph County's loose market.

Even though the evidence comes from only two housing markets, we believe our findings on the price effects of housing allowances would hold in other U.S. housing markets. We base that judgment on the modest increase in marketwide demand caused by the program, on the portability of the allowances which prevents the demand increase from being narrowly focused on specific dwellings or small submarkets, and on the supply responses (repairs, inventory changes, and occupancy change) that together rapidly accommodate the increased demand. Consequently, even during the initial years of a housing allowance program, the suppliers of rental housing would have almost no opportunity to raise prices much above the general price set by nonprogram forces.

With hindsight, the price increase fears of the early 1970s can be seen to have resulted from four misconceptions about housing market behavior. A decade ago many believed (a) that once a household becomes poor it stays poor, so that virtually all eligible households would participate in a housing allowance program; (b) that all poor households live in substandard housing; (c) that once a dwelling becomes substandard it stays substandard; and (d) that the short-run supply of housing services is perfectly inelastic.

Those beliefs logically lead to the conclusion that housing allowances would cause large price increases. A permanent poverty population would quickly join an open enrollment housing allowance program in large numbers; all recipients would have to upgrade from substandard to standard housing causing large increases in their demand for housing; the economic infeasibility of fixing substandard housing would focus that demand increase on a segment of the rental housing market containing affordable standard housing; and the lack of short-run supply response would make price increases carry the entire burden of clearing the market.

However, because of HASE we now know that those beliefs about housing market behavior were inaccurate. First, one-third of households poor today will be nonpoor a year from now, with their place in the poverty population taken by a newly poor household (Carter and Balch, 1981). That poverty dynamic, combined with delays in newly poor households joining a housing allowance program, results in only half of eligible households receiving allowance assistance at any given time, cutting the potential demand shock of a housing allowance program in half (Rydell et al., 1979).

Second, only half of the households that joined the housing allowance program formerly lived in substandard housing at enrollment (McDowell, 1979), primarily because they spent large portions of their incomes on housing (Mulford, 1979). Consequently, only a small portion of a housing allowance payment goes into increased housing consumption; rather, most goes to reduce rent burdens (Rydell and Mulford, 1982).

Third, three-fourths of allowance recipients who originally lived in substandard rental housing repair that housing rather than move to join the allowance program (McDowell, 1979). That flexibility of the housing stock makes over half the rental housing market accessible to allowance recipients and prevents explosive focusing of the program's demand shock (Rydell, 1980).

Finally, especially in loose housing markets, occupancy rates expand in the short run to absorb large parts of demand shifts. Consequently, only small short-run price increases are needed to clear the market while inventory expands to accommodate the increased demand permanently (Rydell, 1979, 1980, 1982).

Hindsight also enables us to see why the Section 8 Existing Housing program caused large price increases (26%) while the housing allowance program did not. In Section 8, landlords tended to raise rents toward the permitted ceiling (the administratively set "fair market rent") when enrolling dwellings in the program, while tenants had no reason to object, since they paid only one-fourth of their income for housing no matter how large the rent. Ironically, the supposedly prudent regulations (a fair market rent ceiling and limitation of allowance payments to actual rent less one-fourth of income) actually caused the price increases they were designed to prevent. In contrast, in the housing allowance program tenants pay the marginal rent dollar, because the allowance depends on the average cost of standard housing rather than on the actual rent of a recipient's dwelling. So tenants are motivated to bargain with their landlords and pay no more than market prices for their housing services.

The policy implication of these findings is clear. Debates about the desirability of housing allowances can center on who participates in the program and how much it benefits them. The program has virtually no effect on those not in the program.

NOTES

1. For a review of the economic reasoning behind the expert judgments and computer simulation models that predicted price increases, see Barnett and Lowry (1979: 2-4).

2. The sum of payments to landlords (contract rent) and direct tenant payments for utilities.

3. Strictly speaking, average rent changes slightly underestimate average price change be-

cause of normal deterioration. We have corrected for that using Follain and Malpezzi's (1980: 94) estimate of normal deterioration rates. They fit hedonic indexes to data from 59 SMSAs. Since the indexes include dwelling age, their estimated coefficients can be used to determine quantity declines over time due to deterioration.

4. The National Bureau of Economic Research housing market simulation model (Kain and Apgar, 1977) and the Urban Institute housing market simulation model (DeLeeuw and Struyk, 1975) predicted large price increases for housing allowance recipients in part because the models make many submarket distinctions. In those models, the additional demand from the housing allowance program gets bottled up in small parts of the housing market, causing large price increases there.

5. We used program data to model the time path of program-induced demand changes, and data mostly from the Annual Housing Survey to model the time path of suppliers' aggregate responses to market signals of excess demand. Then, we estimated the time path of price changes that would be needed to continuously balance the shifting demand for housing services against the available supply. All changes (in demand, supply, and prices) are expressed relative to initial market conditions and abstract entirely from the effects of background inflation in the economy. (For details of model construction and operation, see Rydell et al., 1982.)

6. These and subsequent demand-change estimates in the text are the total changes caused by the allowance program when it has reached its equilibrium size (about four to five years after its start). See Figures 11.1 and 11.2 for the demand changes in the recipient submarket during the program's initial years.

7. Our model assumes that there is no reverse flow of nonrecipients from the recipient submarket to the nonrecipient submarket.

Part IV

Mobility, Neighborhood Change, and Discrimination

□ CHAPTER 12, by Kevin McCarthy, is the first of three papers on the subject of residential mobility and neighborhood change. Unlike conventional housing programs, housing allowance recipients were given the freedom to move from their preprogram dwelling units and neighborhoods. Indeed, this freedom was an important motivation for some people to support and others to oppose the housing allowance concept. McCarthy's chapter reports, using data from the Supply Experiment, that the program had little effect on either the rate at which recipients moved or the geographic pattern of their mobility. This finding raises the possibility that, beyond the constraints imposed by their income levels, poor households face other constraints that hamper their ability to search for and locate better housing.

McCarthy presents findings that indicate that most low-income households utilize the same search procedures with the same effectiveness as other households. The absence of a pronounced mobility effect reflects the relatively low priority that low-income households, working within the constraints imposed by their budgets, attach to housing in general and to moving in particular. Low-income households tend to spend only a small proportion of any additional income they receive for housing. The additional income received in HASE (or in HADE) failed to affect mobility.

Chapter 13, by William Hamilton, focuses on the effect of housing allowances on the economic and racial/ethnic concentration of the areas recipients chose to live. Unlike most forms of federal housing assistance, in a housing allowance program receipt of a subsidy is not tied to a specific dwelling unit. Household choices are constrained by the requirement that units occupied meet certain physical requirements and obviously by the household's other income, and by its family size. Since no upper limit was or need be imposed on recipients' spending for housing, and freedom of choice was possible, early discussion suggested that housing allowances would produce some dispersion of households from their original neighborhoods. These neighborhoods of origin tended to have high proportions of low-income and minority households.

Both Supply Experiment and AAE data indicate that the locational impact of housing allowances was small. The designs of those two experiments made it difficult, however, to distinguish between patterns caused by the housing allowances and those that would have occurred in the absence of a program. Thus, Hamilton's chapter concentrates on the locational choices made by people offered housing allowances as compared to the choices made by a similar population with no housing allowances. The main finding is that recipients of housing allowances make the same locational choices as similar nonrecipients. The chapter then argues that this "noneffect" of an allowance program should be compared with the concentrating effect of most other housing programs. It concludes that the important effect of a housing allowance program in this regard is in the absence of movement toward increased concentration.

Chapter 14, by Carol Hillestad and James McDowell, changes the focus from households to neighborhoods. An allowance program could affect neighborhoods by increasing residents' incomes, thereby allowing additional spending on housing. Further, it could promote repairs or even new construction by conditioning the payment on compliance with the program quality standards. Nonrecipients could also change their behavior; if homeowners and landlords perceive that their neighborhood is likely to improve through an influx of households with higher incomes than before, or through the upgrading of nearby buildings and neighborhoods, nonrecipients might invest in upgrading their properties as well.

Since the Supply Experiment operated in two entire metropolitan areas and enrollment was open to all eligible households, the authors were faced with the difficult task of distinguishing between program-induced effects and changes that would have occurred naturally. Program effects are therefore assessed by comparing EHAP payments as neighborhood inputs with similar, naturally occurring inputs and by comparing neighborhood outcomes by level of program activity. Their study yields little evidence that the allowance program materially altered neighborhood quality—changes in property values and in the physical stock of housing were not systematically related to the level of program activity. (Although no physical evidence of neighborhood change was discovered, recipients as well as nonrecipients perceived positive program effects on their neighborhoods.)

12

Housing Search and Residential Mobility

KEVIN F. McCARTHY

☐ DESPITE EXPECTATIONS to the contrary, the Experimental Housing Allowance Program had little effect on either the rate at which recipients moved or the geographic pattern of their mobility. Since allowance recipients were given both the freedom and the opportunity to move, this finding raises the possibility that beyond the constraints imposed by their incomes, poor households faced other less fully recognized constraints that hampered their ability to search for and locate better accommodations. This chapter uses data collected in the Supply Experiment to examine that possibility.

In general, the findings indicate that although certain households do encounter special problems that hamper their search, most low-income households utilize the same search procedures and with about the same effectiveness as other households. Thus, findings that EHAP had little or no effect on recipients' moving behavior cannot generally be attributed to the special search problems encountered by low-income households. Rather, the low-cost search strategies adopted by most renters in the two Supply Experiment sites appear well-suited to their specific objectives and the recruitment techniques of landlords. Overall, the absence of a pronounced mobility effect reflects the priorities low-income households, working within the constraints imposed by their budgets, attach to housing in general and moving in particular.

In the next section I briefly review the rationale for expecting allowances to affect recipients' mobility and the ways in which the housing search process might diminish those expected effects. I then compare search strategies and the effectiveness with which they are used among different income groups. Next, I examine the question of discrimination in housing search—who experiences it and how it affects the search process. Then I present evidence on landlord behavior, search outcomes, and household preferences that suggests why most renters adopt low-cost search strategies. Finally, implications of the findings are discussed.

RATIONALE

Expectation that the allowance programs would facilitate mobility among recipients were based on the demand-side features of the allowance programs. Unlike traditional supply-side programs which tie subsidies to specific housing units, the allowance program attached the subsidies to recipient households, which, using the allowance payment to supplement their incomes, were then free to select the housing and neighborhoods of their choice. For those enrollees who were dissatisfied with their housing or neighborhood before enrolling, the allowance program provided an opportunity to move. If their housing was substandard, the program actually provided them with an incentive to move, since allowance benefits in the Supply Experiment were restricted to those living in standard housing. Enrollees living in substandard housing were thus forced to choose between making the needed improvement, moving to an acceptable unit, or foregoing the allowance payment.

No one expected, of course, that all recipients would take advantage of these features of the allowance program. Low-income homeowners would be reluctant to pay the sizable transaction costs entailed in selling one home and buying another. Instead, they would be far more likely to repair or improve their current homes than to move to new ones. Eligible minority households, on the other hand, should have been particularly attracted by those features of the allowance program, given the substantial involuntary residential segregation by race characteristic of most American cities (Taeuber and Taeuber, 1965; Sorenson et al., 1975; Schnare, 1977) and the overwhelming preference minority households express for residential integration (Pettigrew, 1973; Taylor, 1979).[1]

Despite these expectations, neither the Supply nor the Demand experiments appear to have changed either the rate or geographic patterns of recipients' mobility. Findings by Hillestad and McDowell (Chapter 14), for example, indicate that the net flows between minority, integrated, and nonminority neighborhoods in St. Joseph County (the one Supply Experiment site with a substantial minority population) were small. Similarly, results from the Demand Experiment suggest virtually no change in the geographic patterns of mobility among either black or white recipients (Atkinson et al., 1980) and little change in the rates of recipients' mobility (Macmillan, 1980).

These findings raise the possibility that in addition to the constraints imposed by their inadequate resources, low-income households face other, less fully recognized constraints that limit their mobility. For example, if income was the principal factor limiting recipients' consumption options, then the extent of housing improvement induced by the allowance program would be limited only by the extent of the subsidy and recipients' preferences for housing vis-à-vis other goods. If, on the other hand, recipients face additional constraints which limit their ability to negotiate successfully for themselves in the market (e.g., they are less adept at searching than other households or are more likely to encounter discrimination), then the allowance subsidies alone may not be sufficient to change recipients' mobility.

There are both conceptual and empirical reasons for assuming that the housing search process could substantially affect recipients' moving behavior. Conceptually, moving is a complex behavior entailing a series of choices rather than a single decision or behavior. These choices, which may not be present in every case, include the decision to consider moving, to undertake an active search, and whether and where to move (Rossi, 1955; Brown and Moore, 1970). Because the search stage intervenes between the decision to consider moving and the actual move, events occurring during the search can play a significant role in determining whether households make their desired adjustments when they move. Empirically, several recent studies (Newman and Duncan, 1979; McCarthy, 1979, 1980; Weinberg et al., 1981) have suggested that the transaction costs of moving—that is, the time, effort, and monetary costs involved in locating and moving to a new residence—are substantial and can significantly affect moving behavior. Moreover, discrimination, both overt and subtle, in the sale or rental of private housing appears to be widespread (Wienk et al., 1979) and can influence both the search and moving decision of those most likely to be affected (Farley et al., 1978, 1979).

SAMPLE AND METHODS

The data used in examining this issue were gathered in the baseline surveys of households and landlords in the two Supply Experiment sites: Brown County and St. Joseph County. The comparisons reported here are restricted to a selected sample of households. This sample includes only renters living in "regular" units (houses and apartments), therefore excluding all homeowners and all occupants of irregular units (mobile homes and rooming houses). Homeowners were excluded because, for reasons already noted, the allowance payments were unlikely to overcome the reluctance of eligible owners to moving.[2] Occupants of irregular units were excluded because their housing circumstances differ substantially from those of other households.

In addition, the comparisons of renters' search procedures and their outcomes are limited to those renters who moved into their residences from other units within the same site in the five years preceding the interview. The search comparisons thus also exclude all inmigrants to each site and all nonmovers, including those who may have searched for a unit but did not move. Inmigrants were excluded because their moving decisions are not directly comparable to those of local movers, since they were not eligible for the allowance program before they entered the two experimental sites. Nonmovers were excluded because the baseline surveys contain no information on their search behavior.

The comparisons reported here focus on differences among households classified in terms of their eligibility for the allowance program rather than in terms of their income per se. The three mutually exclusive categories (referred to as low-, moderate-, and high-income households) are calculated in terms of the difference between one-quarter of a household's adjusted gross income, YAG, and the estimated cost of standard housing, R*, for a household of its size. The three eligibility categories are defined as follows:

Low Income (eligible)	$0.25\ YAG < R^*$
Moderate Income (near eligible)	$R^* \leq 0.25\ YAG < 1.5\ R^*$
High Income (clearly ineligible)	$1.5\ R^* \leq 0.25\ YAG$

Because the Supply Experiment surveys oversampled low-income households, the results must be weighted to represent the population of renters in both sites. However, the comparisons reported here use both weighted and unweighted results. In general, where the comparisons involved regression analysis, the results are left unweighted; where the results represent a simple cross-tabulation or average (whether for households or landlords), they are weighted.

SEARCH STRATEGIES

A household's choice of residence is a significant consumption decision complicated by uncertainty. To reduce the uncertainty about the costs and benefits of alternative actions, households adopt search strategies before embarking on a search. Those strategies necessarily include decisions about what information sources to consult and how much effort to expend, decisions that can influence the type of adjustment they make or whether they move at all. A household's initial plans may, of course, change because of information gathered or problems encountered during the search, and these experiences can also affect search procedures and their outcomes.

If low-income households are less successful at finding units in the market, it could be due to the search strategies they adopt. Although our surveys contain no direct measures of search strategies, they can be gauged in several ways—according to the procedures used in the search. Our focus here is on three measures of the effort expended during the search: (1) the length of that search, (2) the number of units examined, and (3) the number and type of information sources consulted. If low-income households employ different search strategies than other households, they should exhibit differences in each of these dimensions. The evidence bearing on this possibility is reported in Table 12.1, which compares the selected measures of search effort by income level in the two HASE sites.

TABLE 12.1 Comparison of Search Effort Among Active Searchers

	Percentage Distribution by Site and Income Level					
	Brown County			*St. Joseph County*		
Search Characteristic	*Low*	*Moderate*	*High*	*Low*	*Moderate*	*High*
Length of Search						
1 week or less	41.8	40.0	42.0	34.7	36.3	44.0
1-4 weeks	38.4	42.2	42.1	37.2	36.4	31.1
1-3 months	15.6	13.7	14.5	18.7	18.6	20.1
4+ months	4.2	4.1	1.4	9.5	8.8	4.8
Total	100.0	100.0	100.0	100.0	100.0	100.0
Median (days)	12.1	11.3	11.5	16.5	14.5	11.7
Alternatives Examined						
1	38.5	37.4	29.0	33.1	24.8	23.3
2-5	33.4	39.8	39.7	46.9	48.8	48.0
6-11	19.5	18.1	23.0	15.8	17.1	23.8
12+	8.6	4.7	8.3	4.2	9.3	4.9
Total	100.0	100.0	100.0	100.0	100.0	100.0
Median	3.31	3.21	4.08	3.33	3.72	4.13
Percentage Using Source						
Friend or relative	71.1	64.1	64.4	86.4	78.9	79.5
Newspaper ad	83.5	88.1	84.7	76.7	81.9	83.8
Looking at properties	39.7	30.1	30.8	50.1	49.7	44.7
Rental agent	26.8	23.1	25.6	26.5	26.0	32.8
Mean no. of sources	2.21	2.06	2.06	2.40	2.36	2.41

SOURCE: McCarthy (1980: Table 1).

NOTE: Entries are weighted estimates based on a stratified probability sample of 1454 renter households in Brown County and 1114 in St. Joseph County who conducted an active housing search and moved in the five years preceding the surveys.

Focusing on within-site differences, the data provide little support for the hypothesis that low-income searchers exert less effort than their more affluent counterparts. Despite a slightly greater tendency for low-income searchers to choose the first unit they look at, there are few consistent or significant differences among low-, moderate-, and high-income households in terms of search length, number of units examined, or number and types of information sources used. Instead, it appears that most renters, at all income levels, favor a low-cost search strategy. For example, most renters in both sites spend only about two weeks in the market, look at only two or three units besides the one they actually choose, and rely heavily on information supplied by friends and relatives or advertised vacancies in the newspaper.

Even though all classes use the various information sources with approximately the same frequency, they may not use them as effectively. To examine this possibility, I next compare the effectiveness with which low-, moderate-, and high-income searchers use various information sources. Effectiveness, as measured here, indicates the percentage of searchers using each source who

TABLE 12.2 Comparison of Effectiveness of Information Source Usage Among Searchers

	Effectiveness Rate by Site and Income Level					
	Brown County			*St. Joseph County*		
Source of Information	*Low*	*Moderate*	*High*	*Low*	*Moderate*	*High*
Friend or relative	44.8	42.2	45.2	58.4	51.2	49.8
Newspaper advertisement	71.2	73.7	73.2	43.2	52.2	52.1
Looking at properties	11.4	13.8	13.5	17.4	17.4	12.8
Rental agent	7.1	6.1	9.4	.5	6.5	8.7

SOURCE: McCarthy (1979: Table 2).
NOTE: Effectiveness rates are calculated by dividing the number of searchers who located their units through a particular source by the number using that source.

locate their units through that source. If low-income searchers are less adept in using the various sources, that should be evident in differences in effectiveness rates. The data bearing on this possibility are presented in Table 12.2.

The results once again provide little evidence of significant differences in search procedures among the different income groups. In Brown County, for example, the effectiveness rates among all classes of searchers are virtually identical. In St. Joseph County low- and high-income searchers differ slightly in the effectiveness with which they use newspaper and personal referrals. However, these differences are small, and the vast majority of both groups locate their units through one or the other source.

Indeed, the information source usage patterns evident in Tables 12.1 and 12.2, coupled with the fact that contacting friends and reading newspaper listings are generally the least costly ways to find units, support the notion that most renters at all income levels favor a low-cost search strategy. Moreover, given the differential effectiveness of the various information channels, that strategy appears quite reasonable.

DISCRIMINATION AND HOUSING SEARCH

Despite the apparent similarities between their search techniques and those of higher-income households, eligible renters could still be restricted in their ability to use allowance payments to move to better housing and neighborhoods if landlords are reluctant to rent to them because of their racial, ethnic, or other characteristics. Although landlords may have reason to prefer one class of tenant to another, searchers who encounter discrimination will face higher search costs in the form of the additional effort needed to find suitable housing. Moreover, households that expect to encounter discrimination may alter their search procedures—for example, choosing the first available unit or restricting their search to familiar areas—to avoid the humiliation and resentment they might experience. This section compares

TABLE 12.3 Types of Discrimination Encountered

	Frequency of Occurrence (%)			
	Brown County	*St. Joseph County*		
Type of Discrimination[a]	*All Searchers*	*White Searchers*	*Black Searchers*	*All Searchers*
Age	9.2	11.4	13.3	11.6
Sex	5.3	5.1	7.8	5.4
Marital status	13.1	12.7	17.5	13.3
Race	1.2	0.7	20.8	3.7
Nationality	0.8	0.6	11.0	2.1
Income source	7.6	11.6	20.3	12.8
Children	12.2	14.0	22.2	14.8
Pets	12.4	20.1	5.6	17.7

SOURCE: McCarthy (1979: Table 3; 1980: Table 3).

NOTE: Entries are weighted estimates based on a stratified probability sample of 1454 renter households in Brown County and 1114 in St. Joseph County who conducted an active search and moved in the five years preceding the survey.

a. Based on respondents' answers to the question, "While you were searching, was anyone reluctant to rent you a unit because of your (age, marital status, . . .)?

the frequency with which different kinds of searchers experience discrimination and the ways discrimination affects search procedures.

Discrimination is, of course, difficult to measure. One way is to use searcher's responses to questions asking whether anyone was reluctant to rent to them because of their age, sex, marital status, race, nationality or source of income or because they had children or pets. The term "discrimination" is thus used here to denote the number of separate types of discrimination encountered, rather than the number of separate incidents.

Table 12.3 compares the frequency with which all searchers in Brown County, and white and black searchers, respectively, in St. Joseph County, report encountering each of these types of discrimination. Overall, discrimination is more prevalent in St. Joseph than in Brown County—a difference due only in part to the racial composition of the two sites.
For example, while black searchers report higher levels of discrimination than whites in St. Joseph County, even white searchers in that racially diverse community encountered more discrimination than searchers in virtually all-white Brown County. Not surprisingly, very few whites in either site report discrimination on the basis of race or nationality; far more troublesome were marital status, source of income, children, and pets. Blacks, in contrast, report higher levels of discrimination than whites from all sources but one (pets), and over 20% perceive considerable racial discrimination. Overall, these findings suggest that considerable discrimination occurs in both sites and that it is directed at select groups, including minorities, households with children or pets, and families on welfare.

Although reported levels of discrimination appear to be somewhat higher in

TABLE 12.4 Landlord Preferences About Tenant Characteristics

	Landlord Prefers Not Renting to . . . by Site (%)	
Type of Tenant	*Brown County*	*St. Joseph County*
Families with many children	82.8	85.2
Families with pets	78.7	69.9
Unmarried couples	77.1	66.6
Families on welfare	55.9	62.0
Single women with children	53.3	54.7
Families with teenagers	40.5	48.4
Families with young children	34.5	44.8
College students	51.0	40.8
Blacks	32.0	40.6
Single men	44.0	33.4
Single women	29.1	32.7
Elderly	14.1	12.2
Couples without children	4.1	7.5
Whites	0.1	1.3

SOURCE: Tabulations from the baseline surveys of landlords.

the two Demand Experiment sites (Vidal, 1980), that difference is due exclusively to the fact that only low-income households were included in the Demand Experiment, since low-income searchers in Brown and St. Joseph counties report levels of discrimination comparable to those in Pittsburgh. More important, searchers in the Demand Experiment, like those in the Supply Experiment sites, report considerably more discrimination on account of children, income source, and marital status than on account of race or ethnicity.

Tenant reports on the frequency with which they experience various types of discrimination accord closely with landlords' reports on their least preferred tenant types (Table 12.4). For example, over 80% of the landlords in both sites prefer not to rent to tenants with many children, and over half prefer to avoid leasing units to renters with pets, unmarried couples, families on welfare, and single women with children. Somewhat more favored are couples with young children, college students, blacks, and single men and women. On the other hand, few landlords object to elderly tenants, married couples without children, or whites. In light of these reports from both tenants and landlords, it is not surprising that low-income searchers, minorities, and families with children report encountering more discrimination when they search than other households.

The degree to which discrimination restricts searchers' housing options, and thus could account for the allowance programs' failure to change recipients' mobility patterns, will depend on how discrimination affects search procedures. As I have already noted, discrimination could cause searchers to modify their search procedures to avoid discrimination, or it could increase substantially the effort needed to find a suitable unit. Although the Supply

TABLE 12.5 Effects of Discrimination on Search Effort

Number of Discrimination Problems	Search Effort by Income Level and Race: Low-Income Searchers			Moderate- and High-Income Searchers		
	White	*Black*	*Total*	*White*	*Black*	*Total*
	Median Search Length (days)					
None	13.1	13.4	13.1	9.3	12.0	9.5
One	18.8	30.3	21.1	14.1	30.1	14.8
More than one	30.2	36.4	30.5	29.0	30.6	30.2
	Median Number of Alternatives Examined					
None	2.48	2.87	2.58	3.74	2.38	3.57
One	3.54	3.79	3.60	4.23	3.26	4.12
More than one	4.92	4.62	4.87	5.63	6.13	5.83
	Average Number of Sources Used					
None	2.08	2.32	2.13	2.17	2.59	2.21
One	2.42	2.89	2.53	2.45	2.97	2.48
More than one	2.73	3.27	2.89	2.95	2.71	2.91

SOURCE: McCarthy (1980: Table 5).
NOTE: Entries are weighted estimates based on a stratified probability sample of 927 white and 187 black renters in St. Joseph County who conducted an active search and moved in the five years preceding the survey.

Experiment collected no data on how searchers might have modified their search behavior to avoid discrimination, black households in the Demand Experiment reported that they did indeed avoid searching in neighborhoods where they expected to encounter discrimination (Vidal, 1980). Presumably, the same phenomenon may have occurred in the Supply Experiment.

While there is no information on how searchers might have modified their searches to avoid discrimination, one can examine how discrimination affects search efforts. Table 12.5 compares three measures of search effort among eligible and ineligible black and white searchers, controlling for the number of types of discrimination encountered. Since the income level pattern is the same in both sites, the Brown County results are not included here. This comparison highlights three points.[3] First, low-cost search strategies appear to dominate among those searchers who encounter no discrimination. For example, high- and low-income searchers (both whites and blacks) search for less than two weeks, look at two or three units besides the one they eventually choose, and primarily use only two information sources. Second, searchers who experience discrimination, regardless of their race or income, must search longer, examine more alternatives, and use more information channels than those who do not. Third, the effect of different types of discrimination appears to be cumulative, since, with only one exception, each measure of search effort increases with the number of types of discrimination experienced. Thus, the most severe impacts of discrimination will be felt by those households subject to multiple types of discrimination—for example, single black women with children who are supported by welfare and have pets.

Although discrimination can indeed restrict the mobility option of households, its effects are most pronounced only among those who encounter multiple forms of discrimination. Among households encountering a single type of discrimination, its effect was less severe—increasing searches by about a week and the number of alternatives by one. Whether discrimination can account for the allowance program's failure to alter recipient mobility patterns will depend on the frequency with which eligible and ineligible households experience discrimination and, in particular, multiple discrimination. Overall, eligible searchers were more likely to encounter discrimination and to experience more types of discrimination. For example, in Brown County, eligibles were about twice as likely to encounter more than one type of discrimination as ineligibles (20.8% versus 10.0%) but no more likely to experience a single form of discrimination (17.9% versus 17.0%). In St. Joseph County, where discrimination was more prevalent, 28% of the white eligibles (versus 16% of white ineligibles) experienced two or more forms of discrimination, and another 20% (versus an identical 20% of ineligibles) experienced a single type of discrimination. Interestingly, although blacks were more likely to experience discrimination than whites, there was little difference among eligibles and ineligibles in the number of types of discrimination experienced. Approximately one-third of both eligibles and ineligibles encountered multiple types of discrimination, and another 19% of the eligibles (versus 15% of the ineligibles) encountered a single type of discrimination.

Summarizing the evidence, it appears that discrimination may indeed have played a role in the failure of the allowance program to affect some recipients' mobility, since eligibles were more likely to encounter multiple discrimination than ineligibles. However, it seems unlikely that discrimination alone could explain the allowance program's mobility effect; only one-quarter to one-third of eligibles experience multiple discrimination. Over 60% of the eligibles in Brown County and over 50% in St. Joseph County experienced no discrimination at all. Such searchers, both eligible and ineligible, used low-cost search strategies with about the same effectiveness. Moreover, searchers who encountered a single type of discrimination experienced only a moderate increase in search effort.

The HADE results lend further support to this conclusion. Macmillan (1980) reports that although searchers who encounter discrimination must search harder, only infrequently did those difficulties prevent searchers from moving. Indeed, the most frequent reason households gave for terminating their search was their inability to find a unit they could afford—not discrimination.

THE RATIONALE FOR LOW-COST SEARCH STRATEGIES

The apparent preference among renters for a low-cost search strategy is somewhat surprising, given that housing expenses constitute the largest single component of most households' budgets. In addition, since households are not continually in the market for a new residence, and housing, unlike many other consumer products, is a multidimensional good whose attraction for households depends on its particular combination of a wide variety of attributes, we might expect households to search the market more intensely before moving. Consequently, this section considers why most renters, regardless of their incomes, apparently pursue a low-cost search strategy.

Three possible explanations for renter search behavior are considered here: first, the behavior of landlords; second, the outcome of different search procedures; and third, the priority renters place on housing and mobility. Together these results suggest that (1) renters tailor their search procedures to landlord behavior, (2) intensive searches are not particularly effective, and (3) low-cost searches accord well with the priorities renters attach to improving their housing.

LANDLORD BEHAVIOR

In deciding on an appropriate search strategy, renters must consider not only their housing preferences but also the techniques landlords use to recruit tenants. For example, renters would be ill-advised to consult rental agents if landlords rarely use agents to fill vacancies. Indeed, landlord recruiting policies appear to play a major role in influencing how prospective tenants search for housing. For example, most landlords in both sites use only a single technique to recruit tenants—relying either on newspaper advertisements (79% in Brown County and 60% in St. Joseph County) or personal referrals (37% in Brown and 51% in St. Joseph County). Only one other source ("For Rent" signs) was used by as many as 10% of the landlords in either site. Given this pattern of recruitment, it is not surprising that renters in both sites rely so heavily on newspaper ads and personal contacts in looking for units. Nor is it surprising that these are the two most effective information sources.

A clearer picture of why landlords use different marketing approaches emerges when we look at how landlords vary their marketing and management policies by the size of their property. Table 12.6 compares the frequency with which landlords in St. Joseph County rely on one or both of the major recruitment techniques (newspaper advertisements and personal referrals) as well as their general management policies by size of property. These data indicate notable differences in both the recruitment and management policies of owners of different-sized properties. Owners of small prop-

TABLE 12.6 Recruiting Techniques and Management Policies by Size of Properties: St. Joseph County, 1975 (percentage of landlords using technique)

Recruitment Technique/ Management Policy	*Number of Units on Property*					
	1	*2-3*	*4-10*	*11-100*	*101 +*	*All Properties*
Personal referrals	30.6	23.1	12.9	20.8	23.1	27.1
Newspaper ads	34.9	43.9	50.5	16.7	15.4	38.4
Both referrals and ads	23.9	27.6	34.0	52.4	46.2	26.0
Conducts screening checks	50.3	47.6	65.5	66.0	84.6	50.6
Requires leases	26.6	21.5	44.4	50.0	95.0	36.5
Requires deposits	44.7	53.9	76.6	75.0	90.0	49.5

SOURCE: McCarthy (1982: Table 2.3).

erties (1-3 units), for example, are considerably less likely than other owners to use both referrals and newspaper ads and also less likely to conduct screening checks or to require leases or deposits. Owners of the largest properties, on the other hand, typically use both newspaper ads and referrals to recruit tenants and almost routinely require leases and deposits and check prospective tenants' credit references. Finally, owners of mid-sized properties generally fall between these two extremes.

These patterns suggest the importance of the exposure-cost tradeoff in landlords' recruiting and management policies. For the owners of small properties, referrals provide an inexpensive and effective source of tenants, since they have fewer vacancies and correspondingly less need for wide exposure. In contrast, owners of larger properties need wider exposure to attract potential tenants, even though this may mean additional screening checks, leases, and deposits to sort out undesirable tenants. Indeed, all landlords of properties who rely on referrals use significantly fewer screening checks and are substantially less likely to require leases or deposits than are other landlords.

When these patterns are considered in conjunction with the fact that the overwhelming majority of landlords in both sites are small-scale operators, they help explain the informal low-cost search strategies of most renters. For example, over 80% of the landlords in Brown County and nearly 70% in St. Joseph County own only one rental property. Moreover, about two-thirds of the rented units in both sites are located on properties with fewer than five units. Small-scale landlords have neither the time nor the resources for intensive recruitment and are much better served by informal techniques. In St. Joseph County, for example, 70% of the landlords spend less than five hours per week working on their properties; very few employ a management firm (3%) or full or part-time employees (6%); and less than half derive more than

5% of their income from rental properties. In sum, a major factor explaining tenant search procedures appears to be landlord recruiting techniques.

SEARCH PROCEDURES AND THEIR OUTCOMES

Landlords' recruitment and management policies not only help explain tenants' search procedures, they also suggest that tenants who pursue intensive search strategies by using a wide variety of information channels and spending a long time looking at many units may not be rewarded for their efforts. In other words, to the extent that a major proportion of landlords rely on personal acquaintances to find and screen prospective tenants, the outcome of tenants' searches may depend more on who they know than on how hard they search.

There are, of course, several ways to measure the outcomes of a search. Here I focus on the ability of households to find bargains when they move. To determine the effects of search procedures on search outcomes, I regress the dependent variable on variables describing the procedures used in the search and a set of household characteristics.

The household's ability to find a bargain is defined as the difference between the rent actually paid for a given dwelling and the average rent for dwellings with the same attributes. For each searcher's chosen dwelling, I estimate the appropriate average rent by using a hedonic index fitted to the HASE data.[4] The difference between actual and predicted rent is expressed as a percentage of predicted monthly rent—that is, a monthly rent discount. Positive values indicate that households are paying a premium for their units, and negative values indicate that households are getting a bargain.

Seven search variables are used to predict a household's ability to get a bargain: a dummy variable indicating whether a household conducted an active search; the number of units examined during its search; the length of the search; a weighted sum of the information sources used, where the weights are based on the presumed effort involved in their use;[5] the number of types of discrimination encountered during the search; and two interaction terms designed to identify diametrically opposed search strategies.

Table 12.7 shows the regression results for the rent discount equations in Brown and St. Joseph counties.[6] Those results indicate that although search procedures do affect a mover's ability to find a bargain, inside information (as expected) is more important than search effort. In Brown County, for example, none of the three direct measures of search effort (number of units examined, search length, and number of sources used) significantly affected a searcher's ability to find a bargain. Similarly, in St. Joseph County only one of the three direct search effort measures—search length—had the expected effect. In addition, searchers who encountered discrimination, which increased search effort, paid a premium in both sites.

TABLE 12.7 Rent Discount Equations

Variable	Possible Values	Brown County		St. Joseph County	
		Coefficient	*Value of t*	*Coefficient*	*Value of t*
Dependent					
Monthly rent discount (%)	Continuous	—	—	—	—
Search					
Constant	—	3.05		3.12	
No active search	Yes = 1; No = 0	−.10	1.62*	−4.08	.97
Units examined (ln)	Positive Continuous	.86	1.25	.27	.47
Search length (ln)	Positive Continuous	−.27	.78	−.85	1.75*
Sources used	1-10	.11	.46	−.21	.22
Problems encountered (ln)	Positive Continuous	2.88	2.31*	4.17	2.75*
Low intensity–friends	Yes = 1; No = 0	−5.96	3.30*	−4.29	1.53
High intensity–no problems	Yes = 1; No = 0	−1.63	.37	9.80	1.80*
Background					
Local moves	Positive Continuous	−.38	.71	−.53	.98
Length of stay (ln)	Positive Continuous	−.52	1.11	.62	1.02
Head's years of schooling	Positive Continuous	.61	2.40*	.32	1.30
Single male head	Yes = 1; No = 0	1.00	.49	1.85	.74
Single female head	Yes = 1; No = 0	.45	.22	−.39	.14
Single head with children	Yes = 1; No = 0	1.16	.41	−8.20	2.45*
Single-person household	Yes = 1; No = 0	−5.34	2.76*	−4.73	1.95*
Age of Head					
>21	Yes = 1; No = 0	−2.59	.94	−.62	.21
21-29	Yes = 1; No = 0	−6.15	2.98*	1.83	.83
30-39	Yes = 1; No = 0	−.80	.34	−2.24	.88
60-69	Yes = 1; No = 0	−2.73	.79	−3.87	.92

70+	Yes = 1; No = 0	3.61	.81	−3.64	.63
Number of children	Positive Continuous	.75	1.32	2.77	4.07*
Income eligible	Yes = 1; No = 0	−4.71	3.17*	−.75	.42
Near eligible	Yes = 1; No = 0	−1.15	.80	−2.30	1.29
Black household	Yes = 1; No = 0	—	—	.32	.16
Income Sources (%)					
Welfare	Positive Continuous	.05	1.28	−.02	.44
Pensions & social security	Positive Continuous	−.03	.61	−.03	.63
Earnings	Positive Continuous	−.03	.86	−.04	1.30
R^2		.113		.079	
F statistic		4.64		2.91	
Sample size		933		1369	

SOURCE: McCarthy (1980: Table 6).

NOTES: Regression analyses were performed on renters paying full market rents and moving in the five years preceding the baseline survey. A negative sign on a coefficient indicates a discount.

*Coefficient is significantly different from zero at the 0.10 level.

Because the search effort and discrimination measures are correlated, two interaction terms have been used here to identify the effects of distinctly different search strategies. Searchers using the low-cost strategy of searching for a short time and relying exclusively on tips from friends are identified here as inside information searchers; they receive, on average, a 6% monthly discount in Brown County and a 4% monthly discount in St. Joseph County. In contrast, searchers employing a high-intensity search strategy by looking at many units and using all four information sources, and encountering no discrimination, pay average market rents in Brown County and pay a 10% *premium* in St. Joseph County.

The contrasting effects of these two search strategies confirm our expectations that the ability to find a bargain depends more on who you know than how hard you search. This finding reflects the advantages that personal referrals offer both landlords and tenants. Tenants who discover units through tips from friends are able to find bargains with very low search costs. Landlords who rely on referrals to find tenants avoid the cost of advertising and have the additional advantage of being able to screen out unfamiliar and possibly undesirable tenants.

Overall, these results confirm our expectations that intensive searches are generally not cost effective.[7] This, of course, provides additional reason for renters to pursue low-cost, informal search strategies. Moreover, these results suggest that those households that search intensively without encountering discrimination do so not to find a bargain but because they are looking for units that are in short supply.

HOUSING PRIORITIES AND MOBILITY

Since neither discrimination alone nor renters' search patterns appears able to explain why more recipients failed to take advantage of the allowance program's features and change their mobility, we examine one more possibility: the priorities eligible renters attach to improving their housing. As noted earlier, barring significant constraints on their ability to search for and locate suitable housing, the allowance program's impact on recipient's housing improvement would be limited only by the extent of the subsidy and recipient's preferences for housing versus other goods.

Although we have no direct information on recipients' preferences, the surveys did ask respondents questions that can be used to assess their preferences. Specifically, all households were first asked, "If you had $50 more to spend every month, what are the three main things you would do with it over the next year?" Then they were asked, "If you had $50 more every month which you could spend only on housing, would you move, have the landlord improve the unit, or something else?" Finally, those who would move were asked, "Would you stay in this neighborhood or not?" Table 12.8 lists renters' responses to these questions. In both sites, eligible and ineligible renters' responses are listed separately; in St. Joseph County, blacks are listed separately.

TABLE 12.8 Comparison of Renter's Preferences for Spending $50 Additional Monthly Income

	Households Responding (%)					
			St. Joseph County			
	Brown County		*White Renters*		*Black Renters*	
Choice	*Low Income*	*Moderate-High Income*	*Low Income*	*Moderate-High Income*	*Low Income*	*Moderate-High Income*
	How would you spend $50 additional monthly income?					
Housing first choice	12.6	13.5	8.8	10.2	14.8	16.6
Housing among first three choices	23.9	20.3	21.0	21.6	32.5	27.3
	How would you spend $50 additional monthly income on housing?					
Improve current unit	18.7	13.5	21.9	17.6	34.9	25.8
Move to a new unit	50.4	59.2	44.5	52.1	48.9	51.1
Move and change neighborhood	27.2	34.2	30.1	37.2	32.2	38.9
Other[a]	27.9	23.4	27.8	25.7	11.5	10.2

SOURCE: Tabulations from the baseline surveys of households.
a. Includes "Make no change," "Don't know," and other.

To the extent these data provide an accurate measure of renters' preferences, they suggest that the major reason the allowance program produced little change in recipients' mobility is that most renters, both those who were eligible for allowances and those who were not, simply do not attach a high priority to improving their housing. For example, no more than 15% of any group of renters would make housing their first priority if given an additional $50 a month to spend in any way they choose. Indeed, less than one-third of each group of renters ranked housing among their top three priorities.[8] Even if forced to spend the additional monthly amount on housing, only about 50% of the renters would use that money to move to a new unit and only about one-third would move to a new neighborhood. Moreover, eligible renters are actually somewhat less likely both to move and to change neighborhoods. Instead, they are somewhat more likely to ask their landlord to repair their unit, to improve it themselves, or to make no change in their housing whatsoever. Interestingly, black renters seem no more inclined to move or to change neighborhoods than comparable white renters in St. Joseph County.

Two factors could explain these responses. First, housing expenses often strain low-income households' budgets, making them reluctant to use any additional income to increase those expenditures. For example, over 60% of the renters in St. Joseph County who received allowance payments were already spending more than half their adjusted gross incomes on housing before they enrolled. Even after deducting their allowance payments from their housing expenses, two-thirds were still spending more than 25% of their income for housing (Rand Corporation, 1978). This rationale accords closely with findings by Weinberg et al. (1981) that the benefits of moving for low-income households are generally very small, so that relatively large changes in traditional economic variables such as prices and income are likely to have little effect on low-income households' housing demand. Second, in predominantly homeowner housing markets like Brown and St. Joseph counties, renters at all income levels, when given the option, would prefer to own rather than rent—indeed, of the renters in both sites who said they would move, 50% gave as their reason their desire to purchase a single-family house. Consequently, they are reluctant to move and pay higher rents unless their household and housing circumstances dictate. They prefer, instead, to add to their savings until they can purchase.

Regardless of the reasons for these preferences, if they provide any type of accurate indication of recipients' behavior, it is in suggesting that recipients' preferences were a far more important factor explaining mobility behavior than housing search behavior or discrimination.

IMPLICATIONS

Although not definitive, these results are highly suggestive. For example, they imply that demand-side assistance programs with benefit levels comparable to those in the Supply Experiment are unlikely to affect mobility even if coupled with special relocation assistance components such as moving allowances or detailed lists of available vacancies. Furthermore, while strict enforcement of antidiscrimination laws, especially those pertaining to children and minorities, would certainly help some households, they are unlikely to cause major changes in mobility patterns. Instead, these results suggest that the major reason the allowance program failed to affect recipients' mobility was that most recipients were reluctant to spend more than a small portion of their allowance benefits on increased housing, preferring instead to treat the payments as a general income subsidy.

NOTES

1. In the one Supply Experiment site with a substantial minority population, St. Joseph County, over three-quarters of the minority households endorsed the statement that blacks and whites should live in the same neighborhoods.

2. Indeed, homeowner recipients in both sites had very low rates of mobility.

3. The comparison here implicitly assumes that discrimination increases search effort, rather than effort and discrimination being determined jointly. Since an argument can be made for the reverse assumption, we tested for the possibility that the various measures of search effort and number of types of discrimination encountered might be determined jointly. Those tests indicated that although the various measures of search effort are clearly jointly determined (e.g., households that search longer also look at more units), they are not closely correlated with the discrimination measure.

4. A hedonic index consists of a set of housing attributes and associated price coefficients, the latter estimated by regressing rent on attribute values. The coefficients are estimates of the average market price for units of their associated attributes; consequently, multiplying the vector of coefficients by the specified attribute vector of the searcher's chosen dwelling gives the average or "expected" market rent for such a dwelling (see Chapter 9).

5. The weights are as follows: personal contacts = 1, newspapers = 2, driving or walking around =3, rental agents = 4.

6. The household variables are used primarily as controls in these equations. Consequently, we do not discuss their substantive significance. However, they can be interpreted as identifying households for which cost constitutes an especially important factor in choosing a dwelling.

7. Although the comparison here is limited to renters' ability to find bargains, in other papers (McCarthy, 1980, 1982) I have shown that intensive searches have no effect either on the size of the deposits landlords require tenants to pay on moving to a unit or on renters' satisfaction with their dwellings six months after moving.

8. Renters in both sites attached highest priority to saving and paying off bills.

13

Economic and Racial/Ethnic Concentration

WILLIAM L. HAMILTON

☐ UNLIKE MOST previous forms of federal housing assistance to the poor, a housing allowance program offers recipients the opportunity to occupy housing in a location of their own choosing. In the Demand Experiment,[1] this choice was somewhat constrained by requirements that households live within the designated program areas (Allegheny or Maricopa County) and that they occupy rental housing. Some groups were further required to occupy housing meeting specific standards of physical adequacy or specified rent levels. As long as they met these requirements, however, households could choose housing anywhere within a relatively large metropolitan housing market. Even after becoming allowance recipients they could move to new locations, again provided they met the requirements.

The freedom of locational choice with a housing allowance stands in striking contrast to conventional public housing and other construction-oriented programs of subsidized housing, where the major locational decisions are made by the producers rather than the consumers of housing. Advocates of the housing allowance concept considered this freedom a major advantage over conventional programs. For example, when the President's Committee on Urban Housing (the Kaiser Committee) argued in 1968 for the establishment of a housing allowance program, they cited as the first of three "compelling" reasons the fact that "a housing allowance would allow recipient families greater freedom of choice in location and type of housing" (p. 14).

If housing allowances were to be available to a large proportion of the low-income population, this freedom of locational choice could bring about large-scale population redistribution. This possibility produced a great deal of speculation and a bit of empirical research about the impact of housing allowances on residential location patterns. The underlying issue was the possibility that the housing allowances might serve as a mechanism for dispersing existing concentrations of poor people and racial or ethnic minority groups.

Much early discussion of housing allowances assumed that they would produce some dispersion—that low-income and minority households would tend to leave the economic or racial "ghetto" in favor of more heterogeneous neighborhoods.[2] The Kaiser Committee argued for the desirability of such an outcome: "The excessive concentration of people of one narrow income level or age or race in one area should be avoided" (p. 48), and many subsequent commentators took this outcome as a goal. Many opponents of the housing allowance concept viewed the same assumption from a different perspective. Believing that the program would foster locational mobility, they feared the poor and minorities would "invade" the suburbs and accelerate the abandonment of central-city areas.

The early evidence about how a housing allowance program would actually affect locational patterns was ambiguous. The first empirical information came from the demonstration program implemented in Kansas City in 1970. Many housing allowance participants there moved away from the central city into neighborhoods generally considered to be of better quality, and black households tended to move from highly concentrated areas to neighborhoods with a higher percentage of white residents. Solomon and Fenton (1974) describe the Kansas City experience as one of dispersal but also note that most of the movement occurred along already established paths. Hence, the effect of the housing allowance could not be clearly distinguished from the prevailing patterns of mobility.

The empirical evidence from the two other experiments in the Experimental Housing Allowance Program is roughly consistent with the Kansas City experience. Data from the Administrative Agency and Supply experiments suggest that the locational impact of housing allowances is not large. In both experiments, the proportion of participants changing neighborhoods was smaller than that in the Kansas City demonstration. The Supply Experiment indicated that a housing allowance did not contribute much to neighborhood growth or decline; participation was open to all eligible households in the program areas, but the maximum net change in population in any neighborhood was less than 2% of the neighborhood population (Rand Corporation, 1978).

Still, both experiments left often the possibility that a housing allowance might induce some economic or racial deconcentration. The Administrative Agency Experiment found that participants moved, on the average, to neighborhoods characterized by higher income levels than those they left (Abt Associates Inc., 1976). Both experiments found that black households, on average, moved to somewhat less racially concentrated areas than those they started from. As in the Kansas City case, however, the design of the experiments made it impossible to distinguish between patterns caused by the housing allowance and those that would have occurred in the absence of the experimental program.

The Demand Experiment, then, offered the only opportunity to compare the locational choices made by people offered housing allowances to the choices made by an equivalent population with no housing allowance. It showed, in short, that there was no important difference in these choices.

A CROSS-SECTIONAL PERSPECTIVE

In retrospect, one might have expected this outcome. The average housing allowance was less than $60 per month. How much locational change could be expected for $60 per month, especially when people could divide the money as they pleased among improved location, improved dwelling unit characteristics, and nonhousing consumption?

One perspective is provided by a cross-sectional analysis using 1970 census data. Each census tract in the Pittsburgh and Phoenix program areas was coded in terms of its low-income concentration and its minority concentration. Low-income concentration is defined as the percentage of households in the census tract with household income less than $5000 in 1970. Minority concentration measures were the percentage of households in the tract headed by a black person and the percentage with a Hispanic head (Phoenix only). For each measure in each city, two equations depicting the relationships between concentration and household income and concentration and rent were estimated. These equations were then used to predict the amount of changes in concentration that would be expected if (1) rent or (2) household income were to increase by the amount of the housing allowance. The prediction when rent is changed is equivalent to assuming that the household spends all of its allowance to increase housing consumption. The second prediction, in contrast, assumes that the allowance is treated as ordinary income.

These analyses suggest that the housing allowance could be expected to support only a small amount of deconcentration (defined as the extent to which a household's destination census tract has a lower percentage of low-income or minority households than the origin census tract). As shown in Table 13.1, initial concentration levels ranged from 35% to 58% on the various measures. The allowance would not be expected to cause more than a 4-7 percentage point reduction if the *entire* allowance were used to increase housing consumption. Moreover, if participants treated the allowance as ordinary income, they would be expected to average less than one percentage point of deconcentration. Such reductions would not even bring the allowance recipients to the countywide average levels of concentration. To reach concentration levels equal to the county aggregate, participants would require average reductions of more than 15 percentage points in low-income concentration, 30-40 points in black concentration, and 25 points in Hispanic concentration.

The utility of such cross-sectional analyses for predictive purposes is limited. Nonetheless, they do provide a useful a priori indication that the housing allowance probably would have very small effects on patterns of population concentration.

TABLE 13.1 Predicted Changes in Concentration for Households Receiving a Housing Allowance

	Low-Income Concentration		*Black Concentration*		*Hispanic Concentration*
Household Group	*Pittsburgh Households*	*Phoenix Households*	*Pittsburgh Black Households*	*Phoenix Black Households*	*Phoenix Hispanic Households*
Mean initial concentration	35.4%	39.0%	57.6%	42.4%	40.0%
Mean monthly allowance	$49	$66	$50	$64	$71
Predicted change if entire allowance is used for housing	−4.3	−6.6	−4.8	−6.4	−7.2
Predicted change if entire allowance is used as income	−0.6	−0.8	−0.5	−0.7	−0.8
Sample size	(900)	(718)	(205)	(52)	(208)

SOURCE: Atkinson et al. (1980: Tables 2.2, 3.2, and 4.2).

TABLE 13.2 Mean Changes in Concentration (percentage points)

	Low-Income Concentration		*Black Concentration*[a]		*Hispanic Concentration*[b]
	Pittsburgh	*Phoenix*	*Pittsburgh*	*Phoenix*	*Phoenix*
Experimental households:					
mean	−1.1	−2.7	−4.0	−2.6	−4.0
standard deviation	8.1	11.3	23.2	23.8	19.7
sample size	(916)	(7.5)	(211)	(52)	(207)
Control households:					
mean	−1.2	−3.3	2.6	3.1	−4.8
standard deviation	8.2	11.0	16.8	26.3	16.9
sample size	(320)	(282)	(63)	(27)	(69)
Housing allowance effect[c]	0.1	0.6	−6.6	−5.7	0.8

SOURCE: Atkinson et al. (1980: Tables 2.3, 3.4, and 4.3).
a. Black households only.
b. Hispanic households only.
c. Difference between experimental and control mean concentrations.

MEAN CHANGES IN LEVELS OF CONCENTRATION

An initial examination of changes in concentration levels in the Demand Experiment generally conforms to the cross-sectional prediction of little effect. Table 13.2 presents average changes in the level of low-income, black, and Hispanic concentration over a two-year period.

The figures for low-income concentration and Hispanic concentration tell a similar story. Households in both experimental and control groups moved, on average, to neighborhoods that were slightly less concentrated than the neighborhoods they lived in when the experiment began. Members of the control group in each case experienced slightly *more* deconcentration than experimental households, contrary to the starting hypothesis. But the difference was very small and not statistically significant.

The initial story looks somewhat different for black households. Experimental households in both Pittsburgh and Phoenix moved to neighborhoods with slightly less black concentration than their starting locations, on average, while the average change for control households was a small increase in black concentration. The resulting difference in mean changes is in the hypothesized direction, and the magnitude is roughly similar to that predicted in the cross-sectional analysis (were the full allowance used for housing). But the differences are still not statistically significant, due to the substantial variance in concentration levels for all groups.

MULTIVARIATE ANALYSIS OF CHANGES IN CONCENTRATION

Because participating households were randomly assigned to the experimental and control groups, the simple comparison of means should be an

TABLE 13.3 Changed Concentration Levels (in percentage points) Among Households That Moved[a]

	Housing Gap	*Percent of Rent*
Low-income concentration		
Pittsburgh and Phoenix	1.4 †	1.7
	(1.72)	(1.36)
Black concentration		
Pittsburgh	−6.2	−13.8
	(0.76)	(1.32)
Phoenix	−5.2	−1.1
	(0.64)	(0.13)
Hispanic concentration		
Phoenix	0.3	9.6 †
	(0.06)	(1.91)

SOURCE: Atkinson et al. (1980: Tables 2.11, 3.8, 3.9, and 4.8).
† = Significant at the 0.10 level in a two-tailed test.
t-statistic in parentheses.
a. Based on multivariate analyses as described in the text.

unbiased indicator of the treatment effect. But randomization sometimes breaks down on key variables, and attrition in long-term experiments can introduce systematic variation. To take these and other factors into account, a series of multivariate analyses was performed.

The analysis presented here included only those households in the sample that moved at some point during the two years. This allows the clearest view of the housing allowance's effect on neighborhood choice, since any such effect will be manifested only for people who move. Because most households are expected to move sooner or later, this analysis may be somewhat simplistically viewed as an upper-bound estimate of the effect of the housing allowance on deconcentration of the recipient population.[3]

Two estimation procedures were used. For the analysis of low-income concentration, an equation was estimated for control households, with low-income concentration of the household's final location treated as a function of the initial concentration and other household- and neighborhood-level variables. Using the equation estimated for control households, predicted final concentrations were estimated for experimental households, and the analysis focused on a comparison of the mean residuals for experimentals and controls. For the analysis of black and Hispanic concentration, because of the smaller number of cases, an equation was estimated with final concentration treated as a function of treatment group, initial concentration and other household- and neighborhood-level variables. In this analysis, the treatment effect is measured by the coefficient of the treatment group variable.

In order to sort out any differential effects caused by the incentives associated with particular housing allowance plans, the analysis was performed separately for various subgroups of the experimental group. Shown in Table 13.3 are the results for the housing gap and percent of rent plans. The housing gap plan provides the most direct incentive to consume particular kinds of housing standards; the standards do not, however, directly bear on neighborhood characteristics. The percent of rent plan provides the strongest general incentive to increase housing expenditures by making the allowance a specified fraction of rent.

The results of the multivariate analysis do not challenge the impression left by the comparison of means, that the housing allowance had no major effect on low-income or racial ethnic concentration. The direction of the effects shown in Table 13.3 is like that in Table 13.2, with only the analysis of black concentration showing effects in the hypothesized direction. The magnitude of the effect estimates is generally larger than that shown in Table 13.2, owing to the exclusion of households that did not move. None of the housing allowance effects was significant at the 0.05 level. Two effects were significant at the 0.10 level, though—that for the housing gap plan in the analysis of low-income concentration, and that for the percent of rent plan in the analysis of Hispanic concentration—but more detailed analyses provided no reason to treat these as meaningful effects.

Since these near-significant effects lie, in any case, in the reverse direction from that hypothesized, it is clear that the analysis provides no support to the argument that the housing allowance will result in substantial deconcentration.

PATTERNS OF INDIVIDUAL MOVES

Why didn't the housing allowance induce more deconcentration? Part of the answer is provided by the cross-sectional analysis of 1970 census data presented earlier, which shows that subsidies of the magnitude used in the Demand Experiment should not be expected to produce a large effect. Another part of the answer comes from looking at individual households' patterns of locational change. These show enough diversity to suggest either that peoples' preferred levels of concentration differ widely or that their concentration preference is not a dominant factor in neighborhood choice.

Most people who moved during the Demand Experiment did not make a dramatic change in the low-income or racial/ethnic concentration of their neighborhood. This pattern is evident in Table 13.4, which shows changes in low-income concentration for control households. More than half of the households that moved either stayed within their neighborhood or moved to a new neighborhood with very similar low-income concentration (the highlighted diagonal in Table 13.4). Those who did move to substantially different neighborhoods were likely to move "up" to neighborhoods with less low-income concentration, but the pattern was by no means monolithic. Of those households that moved across the category boundaries defined in Table 13.4, one moved to a *more* concentrated neighborhood for every two that moved to a neighborhood with less low-income concentration.

TABLE 13.4 Origin-Destination Transition for Control Movers: Low-Income Concentration

	In Origin Neighborhood				
	0-24%	*25-34%*	*35-49*	*50-100%*	*Total*
In Destination Neighborhood					
0-24%	69%	34%	21%	14%	30%
25-34%	19	52	20	14	20
35-49%	12	14	52	18	24
50-100%	—	—	7	53	11
Sample Size	(77)	(79)	(84)	(49)	(289)
Percentage of all households	27%	27%	29%	17%	

SOURCE: Atkinson et al. (1980: Table 2.7).

TABLE 13.5 Origin-Destination Transition for Black Movers: Black Concentration in Pittsburgh

	Black Concentration in Origin Neighborhood			
	1-14%	*15-49%*	*50-100*	*Total*
Destination Neighborhood:				
0-14%	38%	7%	21%	20%
15-49%	23	74	13	31
50-100%	38	19	66	49
Sample Size	(13)	(27)	(56)	(96)
Percent of all households	14%	28%	58%	

SOURCE: Atkinson et al. (1980: Table 3.6).

Much the same patterns appear when we examine black or Hispanic concentration. Small sample sizes for these analyses make it difficult to draw meaning from the behavior of any subgroup of households. Nonetheless, the general rule is that most households make little or no change in their neighborhood, and those that do make large changes may either increase or decrease their concentration levels.

One interesting point that emerges from the origin-destination analyses is that the "crossflow"—some people moving to increase their concentration while others move to reduce it—is even more pronounced for black and Hispanic concentration than for low-income concentration. Table 13.5 shows, for example, that among those black households in Pittsburgh that made major changes in the level of black concentration, the substantial majority moved to more concentrated rather than less concentrated neighborhoods. The same pattern, though not as strong, holds for black and Hispanic households in Phoenix.

Experimental as well as control households exhibit the relatively low propensity to change neighborhood characteristics and the crossflow of changes in low-income and minority concentration. It does not appear that the housing allowance had any single impact on the pattern (although small sample sizes make this determination difficult). There is no clear tendency for allowance recipients to do more or less changing of their concentration level, or to alter substantially the balance of moves to more concentrated or less concentrated neighborhoods.

IMPLICATIONS

Given the persistently null findings of these analyses, is it fair to conclude that the effect of a housing allowance program on the location of the poor and minorities is a nonissue, or perhaps an issue of ideological hopes and fears without empirical substance?

Not quite. On the one hand, it seems clear that a housing allowance or similar program could not be expected to redistribute much of the nation's low-income population. Analyses described elsewhere in this volume have shown that many people will not even participate in such a program, and many who do participate will stay in the dwellings they already occupy. And the analyses reported here indicate that the allowance recipients who move will choose locations that are largely indistinguishable from what they would have chosen without the subsidy.

On the other hand, it is important to consider the housing allowance's null effect in the context of alternative strategies of housing assistance for the poor. An analysis of public housing, Section 23 and Section 236 programs operating in Pittsburgh and Phoenix at the time of the Demand Experiment indicated that those "comparison" programs tended to increase the concentration of low-income and minority households (Mayo et al., 1980a). The available housing units in the comparison programs were located in neighborhoods with much greater low-income and minority concentration, on average, than the units occupied by either experimental or control households in the Demand Experiment. Participants in the comparison programs tended to come from the neighborhoods much like them, so most participants experienced little change in their neighborhood characteristics. Even so, the comparison programs, especially public housing, tended to move people into more heavily concentrated neighborhoods than they would otherwise have chosen.

The important effect of the housing allowance, then, is not positive movement toward the goal of reduced concentration of the low-income and minority population but the absence of movement toward increased concentrations. It is not the effect many people hoped for and others feared, but it represents an important distinction from the major alternative mechanisms for providing housing assistance to the poor.

NOTES

1. The analysis described here depends principally on data from the Demand Experiment, and in particular from Atkinson, Hamilton, and Myers (1980).

2. See, for example, Netzer (1980), Downs (1973), Peabody (1974), and Weaver (1975). Some of these authors question whether the effect would be dispersion or merely "escape" from existing concentrations followed by the formation of new ones. The (implicit or explicit) assumption that people would flow out of existing concentrations is consistent, however.

3. This would not be true if the housing allowance had substantial effects on mobility rates. Observed differences in mobility, however, were too small to affect these findings.

14

Neighborhood Change

CAROL E. HILLESTAD and JAMES L. McDOWELL

☐ HOUSING ASSISTANCE PROGRAMS often focus on specific neighborhoods requiring rehabilitation or renewal and usually provide assistance for selected dwellings. In contrast, the Experimental Housing Allowance Program was open to low-income households throughout Brown County and St. Joseph County, and recipients were free to move from one dwelling to another within the county without automatically losing their allowance assistance. Two issues to be studied in the experiment were how a program with those features would affect deteriorated neighborhoods and whether existing patterns of racial segregation would be reinforced or weakened.

The allowance program could affect neighborhoods by increasing recipients' incomes, thereby allowing for additional consumption of housing services, and it could prompt housing repairs by requiring participants to comply with the program's housing quality standards. If households perceive that their neighborhood is now or will be a better place to live with an influx of higher-income households or improved conditions in adjacent areas, households and landlords might invest more materials as well as paid and unpaid labor in their properties, thus improving their neighborhoods.

To test the hypothesis that housing allowances alter neighborhoods, we must distinguish program-induced change from other neighborhood change (e.g., the long-standing growth of Brown County). Because enrollees were free to live wherever they chose, it was impossible to designate neighborhoods that were unaffected by the allowance program; and open enrollment precluded having a control group. We assessed program effects by comparing program inputs with similar naturally occurring inputs. We also compared neighborhood outcomes by level of program activity; if there is no clear link between neighborhood change and level of program activity, it was unlikely that the program was an important factor.

PROGRAM ACTIVITY AND NEIGHBORHOOD CHANGE

To measure the spatial distribution of program assistance, each county was divided into small, residentially homogeneous neighborhoods (108 in Brown County and 86 in St. Joseph County). For each site we aggregated the neighborhoods into five neighborhood groupings with comparable numbers of households. Neighborhoods were aggregated on the basis of average cumulative allowance payments per household because that criterion established the clearest association between neighborhood change and the allowance program.[1]

Although the program did not target benefits to particular neighborhoods, as is common in many government assistance programs, the pattern of participation concentrated benefits in certain neighborhoods (see Table 14.1). In neighborhoods with high levels of program activity, average cumulative allowance payments per household were 7 and 14 times as great in Brown and St. Joseph counties, respectively, as they were in neighborhoods with low levels of program activity. The 23 most active neighborhoods in Brown County, containing less than a fifth of all households in the county, jointly received over two-fifths of all allowance payments. In St. Joseph County, nearly half of all payments went to the residents of the 21 most active neighborhoods.

The spatial distribution of program activity differed in the two sites. In Brown County, the incidence of allowance payments was highest in the older urban neighborhoods along the Fox River and parallel to but not on the lakeshore. Participation was low in most suburban neighborhoods but rose in the outlying rural townships. In St. Joseph County participation was highest in central South Bend and on the southwestern fringe of the city, moderate in the older neighborhoods along the St. Joseph River, and low in rural townships.

When neighborhoods' level of program activity is compared with other indicators of neighborhood condition, additional differences emerge. Table 14.2 shows that household incomes and property values were lowest in the neighborhoods receiving the most allowance payments per capita. The indexes of residential dwelling and landscaping quality were lowest in the neighborhoods receiving the most allowance payments. By these criteria, the allowance program benefited the neighborhoods and dwellings, as well as the households, that most needed help.

PROGRAM-INDUCED NEIGHBORHOOD CHANGE

The study yielded little evidence that allowance program materially altered neighborhoods (see Table 14.3). There were subtle changes in the areas' population, demographic characteristics, and housing stock. Incomes increased about 8% per year in both counties; property values rose by 10% per year in Brown County and 7% in St. Joseph County. Neither incomes nor

TABLE 14.1 Cumulative Assistance and Average Household Income, by Level of Program Activity

Neighborhoods Grouped by Average Payments	*Number of Neighborhoods*[a]	*Number of Resident Households (Baseline)*	*Average Cumulative Allowance Payments per Resident Household*[b]	*Percentage of Resident Households Enrolled (After 3 Years)*
		Brown County		
1 (high)	23	8,231	$252	13%
2	21	9,017	151	9
3	14	8,578	97	6
4	20	9,084	55	4
5 (low)	18	8,868	37	2
		St. Joseph County		
1 (high)	21	14,113	$220	21%
2	12	14,780	119	10
3	16	13,886	61	6
4	15	15,786	44	4
5 (low)	18	16,254	16	2

SOURCE: Hillestad and McDowell (1982: Table 4.1).

a. The number of neighborhoods does not add up to the total in each county because some were excluded on the basis of insufficient sample size.

b. Disbursed during the first three program years.

TABLE 14.2 Selected Characteristics of Neighborhoods Grouped by Level of Program Activity: Brown and St. Joseph Counties

Neighborhoods by Program Activity Level	*Annual Income per Household*	*Property Value per Dwelling*	*Relative Quality Rating*[a] *Residential Dwelling*	*Residential Landscaping*
		Brown County		
1 (high)	$ 9,534	$16,141	1.00	1.00
2	10,761	20,862	1.06	1.04
3	12,393	22,005	1.05	1.04
4	14,067	25,320	1.15	1.09
5 (low)	15,330	24,928	1.17	1.12
		St. Joseph County		
1 (high)	$ 8,758	$ 8,613	1.00	1.00
2	10,431	9,266	.99	.99
3	11,566	11,601	1.04	1.03
4	13,264	14,382	1.09	1.10
5 (low)	14,015	20,127	1.11	1.12

SOURCE: Hillestad and McDowell (1982: Tables 4.1,4.2, 4.4).
NOTE: All data refer to neighborhood conditions just before the allowance program began. Data for individual neighborhoods within each group were pooled to calculate the measures shown.
a. Based on observer ratings of residential dwelling and landscaping quality using a scale of 1 (poor) to 4 (good), divided by the average rating for neighborhood group 1.

property values rose more than proportionately in the neighborhoods with the most allowance program activity. The physical quality of the housing stock, as measured by field observation, did not change significantly over this period; and there were no systematic variations between neighborhood groupings. Finally, changes in demographic characteristics, such as the fraction elderly households, single-headed households, and renter households, were not significant and certainly not induced by the program.

There are three possible considerations for the low incidence of program-induced neighborhood change:

- Even though allowance program activity was concentrated in the poorest neighborhoods with the worst housing conditions, the program was not a sufficient stimulus to reverse urban development patterns already established in the two communities prior to the program.
- Allowances added little to the overall household income in a neighborhood. Even in the neighborhood group receiving the most allowance payments, allowances increased neighborhood average incomes by one percentage point or less. Although the program increased the average recipient's income by 20% during his participation in the program, that individual increase is diluted by nonrecipient income, which far outweighs recipient income in all neighborhoods.
- Although program-required repairs improved housing conditions, they were relatively inexpensive and not large or visible enough to encourage nonrecipients to repair. Each program repair averaged $23 in cash costs and $7 in unpaid labor.

TABLE 14.3 Neighborhood Change in Income, Property Values, and Dwelling Quality

Neighborhoods by Program Activity Level	*Average Annual Percentage Change*		
	Household Income	*Property Value per Unit*	*Residential Dwelling Quality*[a]
	Brown County		
1 (high)	5.9%	9.0%	.7
2	8.4	11.4	− .6
3	8.5	10.4	1.3
4	9.4	10.1	.2
5 (low)	8.5	10.0	− .3
	St. Joseph County		
1 (high)	4.0%	3.0%	− .3
2	7.0	5.2	.0
3	7.9	8.1	1.1
4	12.3	3.7	.5
5 (low)	10.6	3.3	−1.2

SOURCE: Hillestad and McDowell (1982: Tables 4.5, 4.6).
NOTE: The average annual percentage change is based on the difference between neighborhood conditions before the allowance program began and after three years of program operations.
a. Based on observer ratings of residential dwelling quality using a scale of 1 (poor) to 4 (good).

> Costs were low because those repairs most commonly remedied health and safety defects in the interior rooms and basements of enrollees' homes.

Despite the natural targeting of allowance assistance, the program's effects on neighborhoods were small. Neighborhoods receiving the most assistance were the poorest neighborhoods with the worst housing—both before the program began and after three years of operation.

PERCEIVED NEIGHBORHOOD CHANGE

In addition to searching for physical evidence of neighborhood change, we examined how recipients and nonrecipients thought their neighborhoods had changed as a result of the allowance program. Attitudes about present and especially future neighborhood conditions have considerable influence on what actually happens. Although we were unable to measure any direct neighborhood change, recipients as well as nonrecipients perceived positive program affects on their neighborhoods.

Table 14.4 shows how St. Joseph County households responded to a series of questions asked after three years of program operation (Brown County results were similar). We separated the responses of recipient and nonrecipient households living in neighborhoods with high levels of program activity (groups 1 and 2) from all households in neighborhoods with low levels of program activity (groups 3, 4, and 5).

TABLE 14.4 Perceived Neighborhood Change by Level of Program Activity: St. Joseph County

	Percentage Distribution of Responses by Level of Program Activity in Respondent's Neighborhood			
	Groups 1 and 2 (High)			
Perceived Program effect	*Recipients*	*Nonrecipients*	*Groups 3-5 (Low)*[a]	*All Neighborhoods*
	Effect of Program on Neighborhood Property Values?			
Increased	67	45	38	45
Decreased	1	6	5	5
No effect	32	49	57	50
	Effect of Program on Neighborhood Property Upkeep?			
Increased	84	50	48	54
Decreased	0	4	3	3
No effect	16	46	49	43
	Effect of Program on Repairs?			
Increased	67	55	40	50
Decreased	1	3	7	4
No effect	32	42	53	46
	Has the Program Affected Your Household?			
A lot	75	3	4	9
Somewhat	15	2	5	5
Very little	4	4	4	4
Not at all	6	91	87	82
	Has the Program Affected Your Neighborhood?			
A lot	29	7	2	5
Somewhat	23	21	5	11
Very little	9	11	10	10
Not at all	39	61	83	74

SOURCE: Hillestad and McDowell (1982: Table 6.1).

NOTE: Entries are based on responses from the 75% of all household heads who were familiar with program details.

a. About 88% of the residents of these neighborhoods were nonrecipients.

Countywide, about half of all households thought the allowance program had caused property values to rise, improved property upkeep, and increased the amount of residential repair. Few households—less than 5%—thought the program had negative effects in those areas. However, the proportion perceiving possible neighborhood effects was substantially greater where the program was active than where it was inactive and was naturally greatest among participants in the active neighborhoods.

More than two-thirds of all recipients in neighborhoods with high levels of program activity thought the program had positive effects on property values, dwelling upkeep and repair, and their own households. More than half of those households thought the program had at least some effect on their neighborhood as a whole.

Nonrecipients living in the same neighborhoods expressed less favorable views of the program. About half of those households reported increased property values, property upkeep, and repairs. But more than 90% saw no overall effect on their households, suggesting that direct program assistance determined whether a household judged itself affected by the program. Over 60% of those nonrecipient respondents living in neighborhoods with high levels of program activity thought the program had no effects on their neighborhoods. Almost 80% of those respondents believed program-related moves had no effect on their neighborhoods.

Households living in neighborhoods that received little program assistance were least positive in their views. Between one-third and one-half thought the program had increased property values, upkeep, and repairs. About 87% claimed the program had no effect on their household, and 88% thought program-related moves had no effect on their neighborhoods.

Comparing nonrecipients in neighborhoods with high levels of program activity with all households (most of whom were nonrecipients) in neighborhoods with low levels of program activity, the former were found more likely to respond favorably and their responses were usually more positive. Since the main difference between those two nonrecipient groups is the level of program activity in their neighborhoods, the positive attitude of nonrecipients living in high-activity neighborhoods is probably a reaction to their neighbors' receiving program assistance and often using that assistance to improve their housing.

In part, these perceptions of neighborhood change can be attributed to what attitude researchers call the "positivity bias"—in this case, a general tendency to respond favorably to questions asked about civic betterment.[2] But the differences between high-activity and low-activity neighborhoods indicate that respondents in the former group were at least aware of program activity even if their households were not directly involved. This suggests that they judged neighborhood conditions to have improved because their neighbor had received assistance.

We doubt that the perceived neighborhood change was prompted by widespread evidence of neighborhood improvement, as we were unable to measure any tangible changes due to the allowance program (see Table 14.3). However, the influence of the program on residents' perceptions, particularly in high-activity neighborhoods, may have forestalled further deterioration and might cause neighborhoods to improve in the future.

RESIDENTIAL MOBILITY AND NEIGHBORHOOD CHANGE

The portability of housing allowances led some observers to speculate that a full-scale program would result in spatial redistribution of low-income households, particularly those belonging to racial minorities. Some hoped and others feared that neighborhoods' economic and racial balance would be

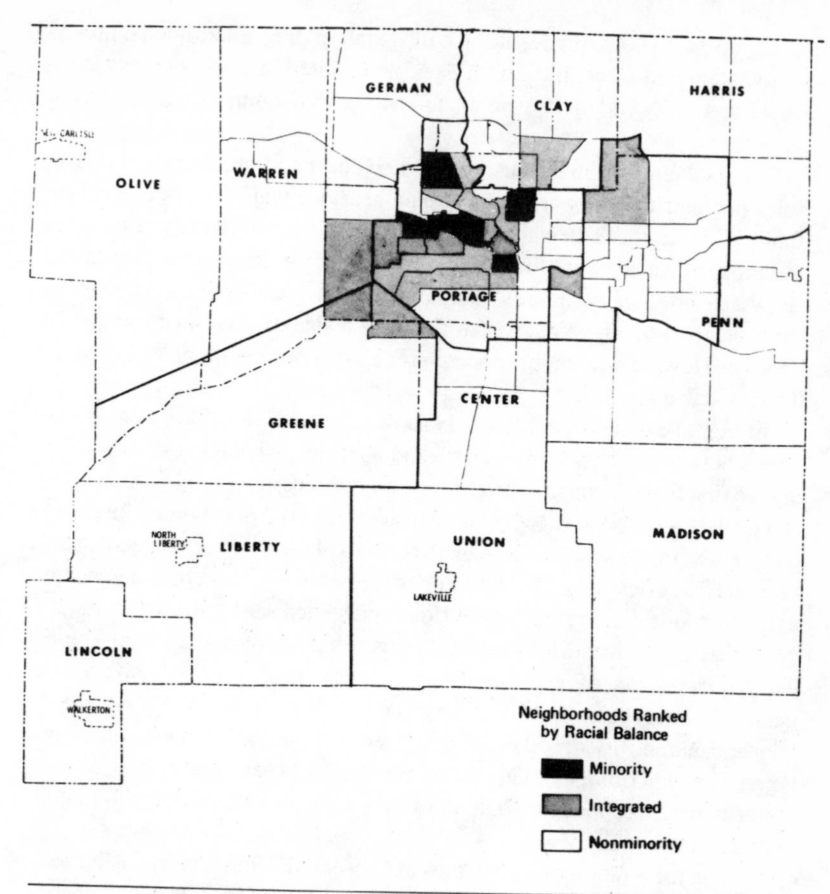

FIGURE 14.1 Racial Balance in St. Joseph County

upset if the program succeeded in removing barriers to economic and racial segregation. One purpose of the research was to learn how many participants moved during the first five program years, what they would gain by moving, and how the neighborhoods of origin and destination would be affected.

We investigated residential mobility and neighborhood change in St. Joseph County, where interest in the issue is sharpened by the fact that a fifth of all allowance recipients were members of a racial minority.[3] To assess the effects of moves on the racial balance of neighborhoods population, each neighborhood in St. Joseph County was assigned to one of three groups:[4]

Minority: 50% or more minority households
Integrated: 5%-49% minority households
Nonminority: less than 5% minority households

Figure 14.1 shows the resulting configuration of neighborhoods: minority or integrated in central South Bend, integrated and nonminority elsewhere in the city and on its fringes, and nearly all nonminority in the rest of the county.

The data confirm that housing and neighborhood problems for households in minority neighborhoods were more severe than those of others (see Table 14.5). In those neighborhoods enrollment was highest and incomes and property values were lowest. The fraction of dwellings predicted to fail allowance program standards[5] was about a third higher in minority than white neighborhoods. Whatever attitudes the residents of minority neighborhoods had toward integration, a search for better housing might well have led them to look elsewhere.

To examine that possibility, we looked at the net flow of program-related moves between minority, integrated, and nonminority neighborhoods. During the first five program years, enrolled households qualifying for payments changed their residences 3641 times, an average of 0.3 moves per household. Only a third of these moves were between two of the three designated neighborhood classifications. The direction of the net flows is away from minority neighborhoods, but the size of the flows, presented schematically in Figure 14.2, was small. Minority neighborhoods lost a net of only 52 households due to five years of program-related moves. Integrated neighborhoods gained 41 households from minority and lost 46 to nonminority neighborhoods. Nonminority neighborhoods gained 57 households, mostly from integrated neighborhoods. The net movement between minority and nonminority neighborhoods amounted to only 15% of the cross-neighborhood moves and 5% of all moves.

Even in the absence of a large flow of households from minority neighborhoods, differences between households moving to and from these neighborhoods may have altered the composition of neighborhood populations. However, we found few socioeconomic differences between households that moved from and to each group of neighborhoods. Because all were low-income program participants, their incomes and housing expenses were similar. There were, however, slight demographic differences. The minority neighborhoods were net exporters (to integrated neighborhoods) of minorities and single parents and net importers of elderly persons. But all such flows were small relative to the aggregate populations of each neighborhood group. Consequently, we conclude that the allowance program did not affect the overall racial and economic distribution of St. Joseph County's population.

SUMMARY

Although housing allowances were not explicitly targeted to particular neighborhoods, the pattern of participation caused payments to be concen-

TABLE 14.5 Selected Characteristics of Neighborhoods Grouped by Racial Composition of Residents: St. Joseph County

Neighborhood Group		*Resident Population*				
Racial Composition[a]	*Number of Neighborhoods*	*Number of Households*	*Percentage Enrolled (year 3)*	*Annual Income per Household*	*Property Value per Dwelling*	*Incidents of Substandard Dwellings*[b]
Minority	6	7,719	32%	$ 7,973	$ 8,104	58%
Integrated	16	21,158	17	8,446	8,856	45
Nonminority	59	45,721	10	10,362	12,825	44
All neighborhoods	81[c]	74,598	14	9,762	11,613	46

SOURCE: Tabulated by authors.

NOTE: All entries except percentage enrolled are based on data from a survey conducted just before the allowance program began. The participation rate is based on the resident population at the end of program year 3.

a. Minority = 50% or more minority households; integrated = 5-49% minority households; nomminority = less than 5% minority households. Nearly all minorities are black; a few are Latinos or Orientals.

b. Estimated from a failure model based on HAO housing standards.

c. Excludes five neighborhoods because of inadequate sample size.

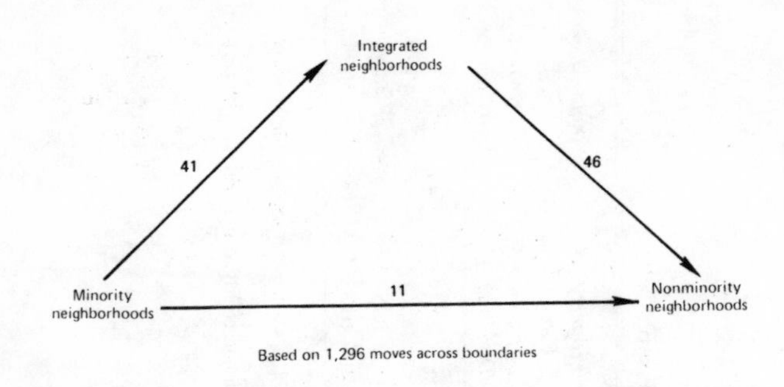

FIGURE 14.2 Net Flows of Movers Between Integrated, Minority, and Nonminority Neighborhoods in St. Joseph County

trated in low-income neighborhoods where housing conditions were worst. However, even in those neighborhoods, the program was not enough to make a substantial difference in average neighborhood income or housing quality. The allowance program, as designed, did have a direct measurable effect on participants and their housing. Expectations that these benefits would indirectly have a large-scale effect on neighborhoods were not met because program effects were diluted by the large number of nonparticipants (almost 93% of all households) and because allowance payments were small in relation to other sources of income (allowances raised average neighborhood income only 1% in the neighborhoods receiving the most assistance).

Community attitudes toward the allowance program were positively influenced by the program's effects on repairs and property upkeep. The perceived change was unlikely to have been prompted by widespread evidence of neighborhood improvement, as no tangible changes due to the allowance program were measured. However, impressions are nonetheless important, particularly in high-activity neighborhoods, insofar as they lead to individual actions that improve the neighborhood.

In short, the allowance program's benefits are virtually limited to participants. It assists them in occupying and maintaining safe, adequate dwellings at costs they can afford. Viewed in those terms, the program was a success. The hope that cash allowances would also serve to stimulate better neighborhood conditions proved unfounded. Quite simply, the stimulus was spread too thin.

NOTES

1. The average cumulative allowance payment per household was based on the total amount disbursed during the first three years of program operations, divided by the total number of households residing in the neighborhood.

2. For more information about HASE research on attitudes, see Ellickson (1978).

3. Although Brown County lacks a segregated housing market, residential mobility was studied in terms of geographic location. Specifically, we compared moves between central Green Bay and the rest of the county. Most of the neighborhoods in central Green Bay belonged to the neighborhood group receiving the most housing allowance assistance. The program-related moves between those two areas were not numerous enough to affect the demographic or economic distribution of Brown County's population.

4. The thresholds were set to provide groupings large enough to include at least 10% of the households. Since minorities are substantially outnumbered by whites in St. Joseph County, there are comparatively few neighborhoods with more than 5% composition, and only a few with over 50% minority composition.

5. In cases where a surveyed household also applied for allowance payments allowance, program evaluation records were matched with household survey records. Whether a unit passed or failed the evaluation was modeled as a function of housing and household characteristics obtained from the household survey records. The model was then used to predict failure ratings for each county.

Part V

Policy Implications

☐ CHAPTERS 1-14 discussed housing allowance programs in isolation. Part V (Chapters 15-17) take a wider view. Chapter 15, by Stephen Mayo, compares program outcomes in three major rental housing programs in existence at the time of the Demand Experiment with outcomes for HADE participants and HADE controls. The three comparison programs are public housing, Section 23 Leased Housing, and Section 236 interest-subsidized housing.

Benefits from the participant's point of view are compared in terms of several measures: (1) the amount of housing provided by each program, as measured by the market rental value estimated using the hedonic indices developed by Merrill (in some of the comparison programs, such as public housing, market rent is not directly observable); (2) the increase in disposable income after rental expenditures are taken into account; (3) the "cash equivalent value" of the subsidy (the estimated unrestricted cash transfer that would make participants indifferent between the housing program and the income transfer); and (4) the participant's subjective measure of satisfaction with housing and neighborhood. Mayo also compares the programs in terms of normative housing standards, including the physical quality of the unit (whether the unit passes HADE's minimum housing standards), crowding, and rent burden. Based on the analysis summarized in this chapter and on a companion analysis that compares program costs, Mayo concludes that demand-oriented programs (such as housing allowances, Section 23 Leased Housing, and Section 8 Existing Housing) are clearly preferable to supply-oriented programs (such as public housing and Section 8 New Construction) in the current economic environment.

Chapter 16, by Marc Bendick and Raymond Struyk, changes the focus from analysis of housing allowances to a discussion of the future of social experimentation. They point out that EHAP was a very costly endeavor. With the benefit of hindsight, they conclude that about $100 million could have been saved if the Supply Experiment had been replaced by computer simulation analyses of housing markets. They point out that computer simulations replicated the actual experimental findings quite well once correct estimates of key parameters relating to the propensity of low-income households to participate in a housing allowance program and their propensity to spend on housing were used. These two estimates were obtained from the Demand Experiment.

Bendick and Struyk point out that the combination of HADE, AAE, and computer simulation could have provided the needed answers more cheaply and more quickly than EHAP. The addition of the Supply Experiment did, however, add face validity to the eventual finding of no price effect, though at a much higher cost. They further suggest that much of the behavioral research that took place in EHAP could have been done with data collected for nonexperimental purposes (e.g., the Annual Housing Surveys). Such research could have been augmented by a limited experiment, such as the Demand Experiment, and a number of demonstrations, such as were run under the AAE.

Finally, Chapter 17, by Edgar Olsen, summarizes what he believes are the implications of EHAP for housing policy. He argues that the findings support two propositions: (1) Low-income people would be better helped if all present housing programs were terminated and the money used for cash transfers; and (2) if there is a special interest in inducing poor families to occupy housing better than the housing they would choose to occupy on their own, current housing programs should be replaced with a universal housing allowance program. The latter point is the position taken by the current administration, though budgetary considerations have not permitted the Department of Housing and Urban Development to propose universal coverage of the low-income population.

15

Benefits from Subsidized Housing

STEPHEN K. MAYO

☐ THE BENEFITS provided by housing programs are multifaceted, from the standpoint of both the program participants and those of the taxpayers who support the programs and the organizations that administer them. A proper reckoning of benefits should have two characteristics: (1) It should be reasonably exhaustive in the multiplicity of program outcomes considered, and (2) it should account for benefits as seen from the viewpoints of both program participants and nonparticipants. This chapter draws on a longer monograph (Mayo et al., 1980a, 1980b)[1] which considers a wide range of comparative outcomes in federal rental housing assistance programs. Among the outcomes considered here are some that may be more heavily weighted by program participants and some more heavily by nonparticipants.

For example, for some concerned nonparticipants, what often appears to matter most about the benefits offered by housing programs is that particular standards be attained, including physical standards of housing quality, occupancy standards, and comparatively recently (as embodied in the Brooke Amendments to the Housing and Community Development Act of 1969) standards concerning the rent level of subsidized households relative to their incomes (sometimes called "rent burden"). Concern with such standards is reflected in program-enabling legislation and in program rules and administrative practices.

The views of nonparticipants do not, of course, provide the only criterion for evaluating benefit levels. Program participants, for example, are more likely to be concerned with satisfying their unique preferences for housing, location, and other household goods and services than satisfying any standard per se (even though housing that meets standards may also be "better" even by the terms of participants).

The evaluation of program benefits in the analysis that follows may be regarded as roughly corresponding to the perspectives of program participants and concerned nonparticipants. The presumed views of program participants are reflected principally in measures of the market rental value of housing provided, the disposable income net of housing expenditures left for participants, alternative measures of economic benefits such as the cash-equivalent value of program benefits, and subjective measures such as expressed satisfaction with housing and neighborhoods. The presumed views of nonparticipants are reflected mainly in measures of the ability of program units to attain alternative standards pertaining to physical housing quality, crowding, and rent burden. While such a dichotomy is admittedly simplistic (achievement of standards may matter to participants and the economic benefits or subjective evaluation by participants of a program may matter to nonparticipants), it is nevertheless useful in suggesting that alternative outcome measures may be complementary and not simply different measures of the same thing.

THE PROGRAMS

The analysis is based on a comparison of program outcomes in three of the major rental housing subsidy programs in existence at the time of the Housing Allowance Demand Experiment with outcomes of Demand Experiment participants in housing gap minimum standards programs. The three programs compared with housing allowances are listed below:

Public Housing (Conventional and Turnkey), comprising low-income housing projects owned and operated by a public housing agency;

Section 23 Leased Existing Housing, comprising housing units from the existing private housing stock leased by a public housing agency;

Section 236 Interest Subsidized Housing with and without rent supplements, owned and operated by organizations in the private sector and comprising housing projects that contain some units for very-low-income households as well as units for moderate-income households.

These programs were selected primarily because they represent the major alternative rental housing assistance strategies that were being pursued by the federal government during the 1970s.[2] Table 15.1, for example, gives the total number of units being provided under each major rental housing assistance program during fiscal year 1974. As the table indicates, Section 236, Owned Public Housing, and Leased Public Housing comprise a substantial majority of all units provided at about the time data were collected for this analysis (1975).

Although the Demand Experiment tested a variety of different housing allowance plans, the analysis presented here is restricted to one type of housing allowance, housing gap minimum standards (HGMS). This is the form of

TABLE 15.1 Unit Breakdown by Legislative Title of Federally Subsidized Rental Housing, Through FY 1974

Program	*Number of Units*[a]	*Percentage of Total*
Section 236	569,910	26
Rent supplement[b]	203,230	9
single subsidy—107,350		
double subsidy— 95,880		
Public housing/owned	1,149,000	52
Public housing/leased (Section 23)	173,700	8
Section 221 (d) (3) BMIR	95,200	4
Section 202	19,700	1
TOTAL	2,210,740	100

SOURCE: U.S. General Accounting Office (1973: Table 4, p. 40).

a. The number of housing units supported through fiscal year 1974.

b. The single subsidy units are those subsidized only by the rent supplement program. Double subsidy units are those subsidized both by the rent supplement program and by one of several other federal subsidy programs, primarily the Section 236, Section 202, and Section 221 (d) (3) BMIR (Below Market Interest Rate) programs.

housing allowance that has been proposed most commonly as a possible national program and is the only housing allowance used in all three major components of the Experimental Housing Allowance Program. HGMS plans provide direct cash payments to recipients, which are conditioned on household size and income and are intended to make up the gap between the cost of modest existing standard housing and the amount a household might reasonably be expected to afford for housing from its own resources (from 15% to 35% of income).[3] In order to qualify for allowance payments, eligible households must live in (or move to) units that meet certain program standards for minimum physical quality and space.

The major differences among the programs analyzed here are in the extent to which they rely on the private market to supply housing and the extent to which they place the responsibility for obtaining decent housing on recipients as opposed to federal and local governmental agencies. In public housing and Section 236, units were newly built or rehabilitated, either under direct contract to local public housing agencies (in the case of public housing) or under regulations administered by the Federal Housing Administration (in the case of Section 236). These units are then offered to eligible households at rents below costs. The extent of the subsidy is usually, but not always, conditioned by income and household size.

The Section 23 leased (existing) housing program and housing allowances, on the other hand, use the existing rental housing stock. Under the original Section 23 program, acceptable units are generally leased from private landlords by a local public housing agency and in turn sublet to eligible households at below-market rents. Under housing allowances, the responsibility for finding and renting acceptable units in the private market rests with recipients. Payments are then made directly to households to help cover their housing costs. The revised Section 23 program and its successor, the Section 8 program (begun after data were collected for this analysis), fall between housing allowances and the original Section 23 program. Under these programs, responsibility for finding acceptable units is generally placed with recipients, but the actual leasing of the unit involves all three parties—the landlord, the tenant, and the public housing agency—with a restriction on the total rent that may be paid.

In the following sections major program outcomes are compared and, when possible, sources of differences (such as differences in program rules and administrative features) are noted. Often it is seen that program outcomes could be made more or less similar by straight-forward changes in program rules. Despite these instances, it is clear that some differences in outcomes are not amenable to simple changes in rules, but are instead the result of fundamental differences in program structure. The next section discusses economic outcomes and subjective outcomes likely to be particularly relevant to program participants. Following that outcomes based on attainment of various program "standards" are discussed that are more fundamental to the concerns of nonparticipants. The final section presents a brief summary.

BENEFITS FROM PARTICIPANTS' POINT OF VIEW

To many, the most apparent benefit from housing programs is the amount of housing participants receive. A common and relatively simple measure of the amount of housing provided in a program is its rental value. While in some programs, such as those that provide leased housing from the private housing stock (e.g., housing allowances and Section 23), rents can be directly observed, for those providing housing not privately traded and with subsidized rents (e.g., public housing and Section 236) private rental values must be estimated.

In this analysis, the market value of housing provided in each program was estimated using a hedonic index of housing developed by Merrill (1980). This index was used to measure the amount of housing being provided in demand-oriented housing programs (housing allowances and Section 23) as well as in supply-oriented programs, since rents paid by program participants in even the former programs could not be clearly judged a priori to be "market" rents.[4] Merrill's index explaining market value was modified to purge influences of tenant characteristics (e.g., length of tenure and kinship with the landlord) from the estimate of market value, such that indices for units in each program are, by construction, estimated rents for units just

TABLE 15.2 Estimated Mean Monthly Rental Values

	Pittsburgh	*Phoenix*
Housing allowances	$129	$164
Section 23	133	151
Public housing (all)	132	158
Section 236 (all)	141	181
Public housing (new in 1975)	157	166
Section 236 (new in 1975)	171	202
Private unsubsidized housing (median)	130	175

SOURCE: Mayo et al. (1980a: Figure 3.2; 1980b: Table 5.1 and Figure 3.1).

coming up for rent in 1975 dollars with landlords and tenants unrelated.[5] Once rents were calculated for units in each program regardless of their age, rents for newly built units in the supply-oriented programs were estimated as well. This latter calculation was done to evaluate the sensitivity of comparative program outcomes to the age distribution of the sampled distribution of units in each supply-oriented program. Were this not done, differences in programs could be found that were merely a function of the length of time over which the program had operated (and its units deteriorated) rather than of inherent differences among programs. Table 15.2 presents estimated mean monthly rental values for units in each program.

As the table indicates, the average value of units in each program is not very different at either site. Section 236 units have the highest average values in each site; other programs have no consistent ranking. At neither site, however, is the average value of Section 236 units more than 20% higher than that of units in the lowest-valued program.[6] In each site the average values of units provided in the subsidized programs bracket rather closely the median value of private unsubsidized housing units, indicating that on average subsidized housing is of moderate quality judged by the standards of the marketplace.[6] Closer examination of the distribution of market rental values in each program indicated that a portion of units in each program were of significantly higher value than the private median (often newer units, as suggested by Table 15.2) but that few units in any program would be likely to be judged of excessively high quality based on their estimated values.

Moreover, it is suggested by the figures in the table that as supply-oriented programs mature (their units become older), initial differences in quality between them and demand-oriented programs tend to lessen. This stresses the importance of evaluating housing program outcomes not simply at a point in time but rather over a program's life cycle. At the low end of the housing quality distribution, proportionately fewer subsidized units in any program have values corresponding to the lowest percentiles of the private rental housing market, indicating that units provided in subsidized programs are overwhelmingly above the lowest-quality levels available in the private market. Thus it appears that quality control mechanisms in each program are generally adequate to exclude poor quality units from the subsidized housing stock.

Another benefit that may accrue to program participants is an increase in disposable income after rental expenses are taken into account. All housing programs have rules that stipulate rent levels to be paid by tenants. In some cases these rules are based solely on a tenant's income; in others, on the cost of providing a unit; in still others, on some combination of tenant income and unit cost. Often such contribution rules have the effect of enabling program participants to rent subsidized housing for less than they might otherwise have spent for housing in the private unsubsidized market.

Accounting for the increase in disposable income associated with program participation requires estimating rent levels that would be paid in the absence of program participation and subtracting "program rents." In this analysis estimated rents in the absence of program participation were based on the parameters of a housing expenditure function estimated for Demand Experiment control households (see Mayo et al., 1980a: chap. 3). The estimated function expresses gross rental expenditures as a function of household income and demographic characteristics such as race, sex, and household size. Such an expenditure function permits one to estimate program impacts not only on income but also on housing consumption. Moreover, it enables one to decompose the sources of program benefits into those arising from increased housing consumption and from increased income available for goods other than housing.

Thus, for example, one may express a naive measure of program benefits, B, as the difference between the market value of housing received in a program, H_p, and the program rent, R_p:

$$B = H_p - R_p \qquad [15.1]$$

This is referred to as a naive measure because it takes no account of welfare losses associated with program-induced distortions in consumption patterns which result in the fact that the cash equivalent value of program benefits may differ from their apparent market value reckoned in terms of the naive benefit measure. This may be decomposed into the sum of a change in the market value of housing resulting from program participation:

$$\Delta H = H_p - H_m \qquad [15.2]$$

where H_m is the estimated market value of housing consumed in the absence of the program; and the change in income available for goods other than housing, which is just equal to the difference between preprogram rent, R_m, and program rent, R_p:

TABLE 15.3 Level and Composition of Tenant Benefits (dollars per month)

	Pittsburgh				Phoenix			
Program	*Total*	*Income Benefits*	*Housing Benefits*	*Sample Size*	*Total*	*Income Benefits*	*Housing Benefits*	*Sample Size*
Public housing	$79	$54	$25	215	113	$72	$41	122
Section 23	52	36	16	82	104	68	36	129
Section 236	28	−3	31	250	72	30	43	80
Housing allowances	77	60	16	80	107	80	27	67

SOURCE: Mayo et al. (1980a: Table 3.5).
NOTE: Totals may not equal the sum of income and housing benefits because of rounding.

$$\Delta Y = R_m - R_p \qquad [15.3]$$

One need only assume that in the private unsubsidized market, rent is equal to the value of housing services ($R_m = H_m$) to see that a naive measure of program benefits is thus:

$$\begin{aligned} B &= \Delta H + \Delta Y \\ &= (H_p - H_m) + (R_m - R_p) \end{aligned} \qquad [15.4]$$

which reduces exactly to equation 15.1. Table 15.3 presents estimates of the naive benefit measure and its decomposition into "housing benefits," ΔH, and "income benefits," ΔY, for each housing program.

The table indicates that despite providing significant benefits through housing changes, the predominant means by which three out of four programs (except Section 236) convey benefits to program participants is through changes in income for goods other than housing. It appears, for example, that for programs other than Section 236, from roughly two-thirds to three-quarters of tenant benefits are in the form of income rather than housing.[7]

Section 236, which provides a generally lower level of benefits than the other programs, provides a relatively greater proportion of benefits in the form of expected increases in housing consumption. In Pittsburgh, Section 236 benefits are estimated to result entirely from housing changes; there is in fact evidence that disposable incomes of participants may be lower than those of similar unsubsidized households as a result of program participation.[8] In Phoenix roughly 40% of Section 236 benefits appear to result from increases in disposable income after rent—a much lower fraction than that provided by other programs there.

Another way to measure tenant benefits is in terms of the cash equivalent value of the subsidy—the estimated unrestricted cash transfer that would make a program participant just as well off as accepting a program offer of housing at a subsidized rent. This measure is based on an important insight regarding in-kind transfers such as those represented by housing programs; in general, transfers in kind will be perceived by recipients as worth less than the cost (or the market value) of providing those transfers. The difference between the cost of the subsidy and its cash equivalent value represents a "deadweight loss"—the program costs in excess of the minimum required to bring about a given level of well-being among program participants.[9]

In this analysis the cash equivalent value of the subsidy in each program has been calculated as a Marshallian measure of consumer's surplus. The measure calculated was based on a log-linear demand (expenditure) function such as that described above which was used to estimate consumption in the absence of program participation. In mathematical terms the cash equivalent value based on a Marshallian measure of consumer's surplus embodying a log-linear demand function is:[10]

$$CEV = \left(\frac{1}{H_m}\right)^{1/b} \left(\frac{b}{b+1}\right) \left[H_p^{\frac{b+1}{b}} - H_m^{\frac{b+1}{b}} \right] + R_m - R_p \quad [15.5]$$

where CEV = cash equivalent value,
H_m = predicted housing consumption in the absence of the program,
H_p = housing consumption for program participants,
R_m = estimated rent in the absence of the program,
R_p = actual rent (subsidized) for program participants, and
b = price elasticity of demand.

The Marshallian measure of cash equivalent value may be thought of as composed of two parts. The first, comprising the terms in parentheses and brackets, depends on the amount of extra housing provided by a program; that is, on the terms H_p and H_m. The second is simply the additional disposable income brought about by paying a rent R_p in a program rather than a rent (usually higher), R_m, in the absence of a program. Decomposed in this way, it is seen that the cash equivalent value is not very different from the naive measure of benefits presented earlier. Indeed, both measures of benefits have, in effect, the common term $R_m - R_p$, which is the change in disposable income resulting from program participation. But whereas in the simple benefit measure an extra dollar of housing is counted as being worth exactly a dollar by program participants, in the calculations of cash equivalent values extra housing is discounted based on a household's relative preference for housing vis-à-vis other goods.

Table 15.4 indicates that cash equivalent values are, as expected, lower than the measure of tenant benefits presented earlier (Table 15.3). On the other hand, estimated deadweight losses appear to be relatively modest and somewhat smaller than some previous estimates. For example, the largest estimated median deadweight loss as a proportion of the subsidy is for Section 236 in each site—25% and 24% in Pittsburgh and Phoenix, respectively. Deadweight losses for Section 236 without rent supplements were estimated to be 29% of the subsidy by the National Housing Policy Review (U.S. Department of Housing and Urban Development, 1973: 4-19). Losses estimated here for public housing, 13% and 17% in Pittsburgh and Phoenix, respectively, are correspondingly lower than the estimate of the National Housing Policy Review, 25%. By contrast, the estimated deadweight loss as a proportion of the subsidy for housing allowances is only 9% in Pittsburgh and 12% in Phoenix. The comparatively modest losses in efficiency that are indicated here for most programs are a direct concomitant of the predominant compo-

TABLE 15.4 Estimated Cash Equivalent Value and Deadweight Loss of Subsidy (dollars per month)[a]

	Pittsburgh			Phoenix		
Program	Cash Equivalent Value	Deadweight Loss	Median Loss as Proportion of Subsidy[b]	Cash Equivalent Value	Deadweight Loss	Median Loss as Proportion of Subsidy[b]
Public housing	$67 (27)	$12 (10)	0.13	$91 (31)	$22 (17)	0.17
Section 23	45 (28)	7 (8)	0.07	77 (42)	26 (28)	0.18
Section 236	13 (41)	15 (12)	0.25	52 (40)	21 (16)	0.24
Housing allowances	67 (38)	10 (10)	0.09	86 (58)	21 (31)	0.12

SOURCE: Mayo et al. (1980a: Table 3.10).
NOTE: Standard deviations in parentheses.
a. Cash equivalent values are calculated according to a Marshallian measure of consumer's surplus with a price elasticity of demand equal to −0.22.
b. The subsidy is defined as the market value of participants' housing less the rent they pay for it.

TABLE 15.5 Ratings of Dwelling Unit and Neighborhood Satisfaction (deviation from control group score)[a]

	Pittsburgh		*Phoenix*	
	Dwelling Unit	*Neighborhood*	*Dwelling Unit*	*Neighborhood*
Housing allowances	0.4**	0.0	0.2	0.1
Public housing	0.0	−0.5**	0.2*	−0.5**
Section 23	−0.1	−0.5**	0.2*	0.0
Section 236	0.0	−0.3**	0.3**	−0.2†
Mean control group score	3.1	3.3	3.4	3.5

SOURCE: Mayo et al. (1980a: Table 6.1).
a. Scores were originally reported are based on the following scale: very satisfied = 4; somewhat satisfied = 3; somewhat dissatisfied = 2; very dissatisfied = 1.
**Significantly different from controls at 0.01 level.
*Significantly different from controls at 0.05 level.
†Significantly different from controls at 0.10 level.

sition of tenant benefits in terms of income changes rather than housing changes.[11] Moreover, since housing changes relative to controls are estimated to be relatively modest, any housing "distortion" induced by the programs analyzed here is also relatively modest—also resulting in comparatively small losses in transfer efficiency.

It may even be the case that deadweight losses are smaller on average than those calculated here. As discussed in Mayo et al. (1980a), reliance on a Marshallian rather than a Hicksian benefit measure, using what may be an artificially low price elasticity of demand estimate, and not having controlled for selection biases in evaluating program impacts, may all contribute to inflating estimates of deadweight losses. For example, assuming a price elasticity twice that used here (which is more in line with the literature)[12] would reduce the estimated deadweight loss as a proportion of the subsidy by roughly half.

Thus it appears that while inefficiencies in consumption by housing programs may be less than previously thought,[13] changes in housing induced by such programs may also be less. It is apparent, therefore, that not only housing allowances but also some other housing programs may be thought of more as income transfer programs than as housing transfer programs.

A final outcome of interest to housing program participants is how they feel about the package of housing and neighborhood attributes with which they are provided. In this analysis, households were asked to give opinions of their dwelling units and neighborhoods in terms of satisfaction, ranked on a four-point scale from "very satisfied" (ranked 4) to "very dissatisfied" (ranked 1). Table 15.5 illustrates average control group ratings for dwelling unit and neighborhood satisfaction and deviations from the control group for each housing program.

The table illustrates the following:

(1) Both the control group and subsidized housing participants are on average satisfied with their dwelling units. Mean levels of expressed satisfaction range between "somewhat satisfied" and "very satisfied."
(2) Subsidized housing program participants generally express no lower levels of satisfaction with their dwellings than do control households; housing allowance participants in Pittsburgh and participants in the three programs other than housing allowances in Phoenix express significantly higher levels of satisfaction than do controls.
(3) Subsidized housing program participants are frequently significantly less satisfied with their neighborhoods than are control households. The only exceptions are in the case of housing allowance participants at each site and Section 23 participants in Phoenix, who express levels of neighborhood satisfaction no different than control households.

The latter finding is perhaps most striking—and the most revealing of a major difference in the structure of different housing programs. That is, the programs examined here permitted varying degrees of choice in the housing and neighborhoods they offered potential participants. Along a continuum, housing allowances offered the greatest freedom of choice, permitting qualifying households either to stay in their existing units or to move to a new unit of their choice. Section 23, which leased units from the existing housing stock presented an intermediate position, in general, enabling households to choose from among a geographically diverse number of units.[14] At the other end of the continuum are the supply-oriented programs, Section 236 and public housing, which offer only a limited number of neighborhood choices to participants—often on a "take it or leave it" basis. These differences in the degree to which program choices are restricted should be expected to exert a major influence on subjective program outcomes and indeed appear to have exactly the impact one would expect. Thus the benefits that are provided by several programs in the form of increased housing or income seem certain to be partially offset by restrictions imposed on neighborhood choice; this is most acute in the case of Section 236 and public housing but appears unequivocally not to be the case for housing allowances.

BENEFITS IN TERMS OF ATTAINING HOUSING "STANDARDS"

A variety of measures have typically been used as yardsticks against which to measure the adequacy of the housing stock in the United States—the ability to pass standards of physical quality, crowding, and the relationship of rent or housing cost to income (rent burden). Such measures may also be used to gauge the comparative performance of subsidized housing programs. These normative measures represent a necessary and desirable adjunct to the "economic" measures of housing outcomes discussed in the previous section in characterizing the housing provided in each program. For example, the market values of units provided by different programs may not agree with the values assigned to units by either their tenants or policymakers. Dwelling unit features that are positively valued by the private market may have little value to some subsidized tenants and less, even negative, value to the collective providers of subsidized housing. (So-called "luxury" features of housing, which are valued positively by the market, may be evaluated negatively by taxpayers if incorporated into subsidized housing.) On the other hand, features such as those directly affecting the health or safety of tenants may not be adequately valued by the market or even by the tenants themselves. As a result, evaluation of normative measures as well as of measures based on market value is necessary and important.

The analysis that follows compares a number of measures of physical quality, crowding, and rent burden for the comparison programs and housing allowances, and also compares outcomes for program participants to those of similar unsubsidized households. The latter comparisons help to illuminate the major ways in which subsidized programs may benefit their participants.

PHYSICAL QUALITY OF HOUSING

Subsidized housing programs are established in part to provide safe and adequate housing and to improve the housing quality of participating households relative to what those households would otherwise have experienced. Different programs have established different standards and different mechanisms to guarantee that standards of physical adequacy are met. Programs that rely primarily on new construction, such as public housing and Section 236, have detailed minimum property standards that specify construction materials and techniques that must be used. Programs that rely on the existing stock of housing, such as Section 23 and housing allowances, have standards that specify acceptable and unacceptable dwelling unit characteristics. Other program features, such as cost limits and inspection and maintenance policies, all of which may differ among programs, further influence the level of housing quality.

The analysis presented here is based on alternative measures that were used within the Demand Experiment to define standard housing. For example, one measure of physical quality is based on the program minimum standards used in the Demand Experiment as a criterion for determining whether units occupied by participants in housing gap minimum standards allowance programs would or would not entitle participants to receive a subsidy. That standard, which is based on an item-by-item evaluation of dwelling unit features, is modeled on habitability codes proposed by the American Public Health Association and the Public Health Service. An alternative, more restrictive, standard (referred to as minimum standards high) is also used.[15]

TABLE 15.6 Percentage of Units That Pass Alternative Housing Quality Standards[a]

	Pittsburgh			*Phoenix*		
	Minimum Standards			*Minimum Standards*		
Program	*Program*	*High*	*Sample Size*	*Program*	*High*	*Sample Size*
Housing allowances	74	10	89	86	44	89
Public housing	64	50	241	75	50	225
Section 23	33	6	94	48	30	145
Section 236	63	47	281	63	46	87
Controls	39	11	307	36	20	274

SOURCE: Mayo et al. (1980a: Figures 4.1 and 4.2).
a. Pass rates for housing allowances and controls are calculated at the end of two years of the Demand Experiment.

Table 15.6 presents rates at which units in each program pass alternative housing standards. Major findings illustrated by the table include the following:

(1) Participants in both comparison programs and housing allowances generally pass the Demand Experiment program standard at rates significantly greater than rates for control households. The only exception is in the case of Section 23 in Pittsburgh.
(2) Housing allowance units have the highest estimated pass rates using the program standard—not a surprising conclusion in view of the requirement that housing allowance participants' units meet the standard initially (but not subsequently) in order for households to receive subsidy payments.
(3) When the more rigorous standard, minimum standards high, is used to compare outcomes, both public housing and Section 236 units have higher pass rates than do housing allowance units at each site. This suggests that having satisfied the explicit program requirement, there appears to be only modest pressure or inducement for housing allowance households to further increase housing quality—at least in the quality dimensions represented by the high standard. As a result, for housing allowance households there are sharp increases in failure rates in going from the program standard to the high standard. Apparent differences between housing allowance units and control households units are much smaller based on the high standard than the program standard.

These findings suggest that while both housing allowance and the comparison programs generally offer housing that is better than that occupied by control households, the dimensions in which housing quality improvements come about may be limited rather strictly by the specific housing quality standards of each program. Thus while units in each program may meet the housing quality standards of their own program, there is little guarantee that they will meet those of other programs or even standards represented by slight variations on the standards of their own program. Considerable caution is suggested in comparing housing quality based on the ability of units to attain specific, and arbitrary, housing standards.

CROWDING

Crowding is another housing outcome of interest to policymakers. The degree of crowding in each program is influenced largely by rules or guidelines that govern the permissible number of persons for a given amount of dwelling unit space. Rules for each of the programs evaluated here are posed specifically in terms of the number of persons per bedroom. In the case of housing allowances these rules are absolute, in that households must comply with program occupancy rules in order to receive a subsidy. For the other programs, the rules are more often used as guidelines, with occasional exceptions permitted.

The three comparison programs are alike, in that both minimum and maximum numbers of household members are specified for each size unit. Such rules generally govern the assignment of households of particular sizes to units of particular sizes. Housing gap minimum standards allowances specify occupancy standards only in terms of the maximum number of persons per bedroom and also specify that bedrooms must be "adequate" based on the physical standards imposed under that program. In contrast to the diverse program rules governing the physical quality of units in each program, rules governing occupancy and, hence, crowding are strikingly similar among programs. Suggested upper limits in each program are generally about two persons per bedroom. As a result (depending on the measure of crowding used), crowding outcomes may be expected to be more similar among programs than is the case with regard to physical quality. As in the case of physical quality, however, policies regarding inspection and enforcement of program rules may affect actual outcomes. Information on comparative inspection and enforcement policies was not obtained, so it is not possible to hypothesize about relative program impacts.

Crowding is typically measured in two ways—as the ratio of the number of persons inhabiting a dwelling unit to a measure of the space the dwelling contains, and in terms of whether a household exceeds some crowding standard also measured in terms of the ratio of persons to space. In this analysis, two measures of crowding are used: (1) the number of persons per bedroom and (2) the percentage of households with more than two persons per bedroom. Table 15.7 illustrates crowding outcomes in each program.

The data indicate that on average, all programs have fewer than two persons per bedroom (roughly the program standard specified by each program), and thus their units are not crowded overall. Levels of crowding are generally similar for the comparison program households and for housing allowance households at each site; in each case overall ratios of persons per bedroom are below

TABLE 15.7 Crowding Outcomes[a]

Program	*Average Number of Persons per Bedroom*		*Sample Size*	*Percentage with More than Two Persons per Bedroom*		*Sample Size*
	Pittsburgh	*Phoenix*		*Pittsburgh*	*Phoenix*	
Housing allowances	1.49	1.33	89	11	5	89
Public housing	1.26	1.55	241	3	7	225
Section 23	1.76	1.23	94	25	6	145
Section 236	1.26	1.21	281	3	1	87
Controls	1.94	1.90	307	26	22	273

SOURCE: Mayo et al. (1980a: Table 4.9).
a. Outcomes for housing allowances and controls are calculated at the end of two years of the Demand Experiment.

corresponding ratios for control households. Each program nevertheless has some households that are crowded, although proportions crowded in both the combined comparison programs and housing allowances are generally well below corresponding proportions among control households. Among individual comparison programs, Section 236 is least crowded based on each measure at each site. That fact, as well as other comparative outcomes on crowding, depend heavily on the size distribution of households among programs. While the results are not presented here, in general, the larger the average household size in a program, the more crowding there is in the program based on the crowding measures used here. Controlling for differences among programs in average participant household size reduces the apparent differences among programs illustrated in the table, such that all appear to have roughly similar impacts in reducing crowding.

To summarize the analysis of crowding, it appears that subsidized housing provided by the comparison programs or housing allowances is on average uncrowded, is less crowded than the housing occupied by control households, and has a relatively greater impact in reducing crowding for large households than for small ones. Comparative uniformity in occupancy rules among programs is reflected in similar numbers of persons per bedroom for similar-sized households in all programs.

RENT BURDEN—THE RELATIONSHIP OF TENANT RENTS AND INCOMES

An important goal of U.S. housing policy that has emerged relatively recently is that households be able to occupy their housing at a reasonable cost relative to their means. A major analysis by the Joint Center for Urban Studies (Birch et al., 1973) was one of the first major evaluations of recent U.S. housing problems to cite a growing crisis in the affordability of housing. In that analysis it was noted that while the incidence of substandard and overcrowded housing had fallen consistently during the twentieth century, during recent years the fraction of the population spending in excess of the commonly perceived norm of 25% of their incomes on housing had grown.

The perception of the impending crisis among the general populace was preceded by recognition of a more severe manifestation of the same problem among publicly subsidized housing tenants. Public housing, until the late 1960s, required tenant rental payments to cover project operating costs. As operating costs grew during the post-World War II period, tenants' rental schedules were shifted upward again and again, more rapidly than public housing tenants' incomes had increased.[16] As a result, by the mid-1960s, tenants in public housing often paid rents well in excess of 25% of their income. The perception grew that while public housing may have been providing decent housing for those who otherwise would not have been able to obtain it, it was doing so at the cost of impoverishing its tenant population. The legislative response to the problem was the Brooke Amendment to the Housing Act of 1969, which set a maximum on tenant contribution rates and simultaneously established federal operating cost subsidies to public housing. The latter enabled tenants to pay no more than 25% of their adjusted incomes for rent. The Brooke Amendment, more than any other factor, focused national attention on the problem of rent burden and the affordability of housing. Furthermore, it boosted the perceived legitimacy of the conventional rule of thumb, that households should spend no more than 25% of their incomes for rent.

Ways in which tenant rents are established and the relationship they bear to tenant incomes vary greatly among programs, however. In public housing and Section 23, the Brooke Amendment rules apply and thus rents are based on tenants' adjusted incomes and are limited in principle to 25% of adjusted income. In Section 236 there are two different sets of rules—households that receive rent supplements pay rents that depend principally on their incomes, with contributions generally limited to 25% of adjusted income, and households that do not receive rent supplements pay rents that are related to the construction costs of their units, with no explicit relationship between rents and income. For recipients of housing allowances, the relationship between net tenant contributions for rent and incomes depends both on the gross rent the household chooses to spend and on the housing allowance subsidy payment; while the latter is related to income by an explicit program rule, the household is free to spend what it chooses for gross rent. Thus the net rent levels for housing allowance recipients bear no fixed relationship to their incomes.

Rent burdens as defined in this analysis are calculated by dividing monthly tenant contributions for rent, including utilities, by net monthly income, as it is defined for most analytical purposes in the Demand Experi-

TABLE 15.8 Median Rent Burdens and Percentages of Households with Rent Burdens in Excess of 25% (percentages)

	Pittsburgh			*Phoenix*		
Program	*Median Rent Burden*	*Percentage with Burden in Excess of 25%*	*Sample Size*	*Median Rent Burden*	*Percentange with Burden in Excess of 25%*	*Sample Size*
Public housing	19.6	13.4	253	20.6	22.6	217
Section 23	16.3	8.6	93	24.7	45.1	144
Section 236						
with rent supplements	22.8	34.8	222	22.1	29.0	59
without rent supplements	33.3	74.3	66	37.5	89.3	31
Housing allowances	20.5	31.7	82	21.4	37.6	85
Controls	27.4	56.0	291	31.1	71.6	236

SOURCE: Mayo et al. (1980a: Table 4.11).

ment. Common definitions of rent and income were used across programs in order to facilitate comparisons. Table 15.8 presents the median rent burden and the proportion of households with rent burdens in excess of 25% of their incomes for each program. Section 236 results are presented for both rent supplement and non-rent supplement households because of their very different tenant contribution rules.

The table indicates a clear difference between Section 236 without rent supplements and all other programs, a result largely attributable to program rule differences. Because Section 236 participants who do not receive rent supplements most frequently pay a cost-based rent on their dwellings rather than a rent that is related to their incomes, their median rent burdens are above 35% at each site, and the proportion of households paying greater than 25% of income for rent is 74% and 89% in Pittsburgh and Phoenix, respectively. At each site the median rent burden for Section 236 households without rent supplements is above the level for control households, as is the proportion paying in excess of 25% of income.

For programs other than Section 236 without rent supplements, the median rent burden is below the level of control households' level and below the level of policy concern, but nevertheless some households in each program are paying rents in excess of 25% of income. In Pittsburgh, the proportion with excessive rent burdens ranges from about 9% for Section 23 to about 35% for Section 236 with rent supplements; in Phoenix, the range is from about 23% for public housing to about 38% for housing allowances.[17]

Differences in rent burden between program participants and controls are consistent with the sizable income benefits in most programs noted in the previous section. Comparisons of participants and control outcomes suggest that for programs other than Section 236, increases in income resulting from reduced rent burdens are on the order of from roughly 6% to 11% of household income.

SUMMARY AND CONCLUSIONS

While the benefit measures discussed here are far from an exhaustive reckoning of the range of benefits to either housing program participants or society at large, they do convey a considerable sense of major program benefits and of systematic similarities and differences among programs.

The housing provided in the subsidy programs examined here, for example, appears to be better than that occupied by nonparticipant control households. Given the focus of the programs studied on low to moderate income groups, it is not surprising that the housing provided in each program is of rather modest quality—with average estimated private market rental values at about the level of median unsubsidized rents in each city. Differences in housing offered in the subsidy programs and that occupied by controls are not in general large—roughly on the order of 15%-20% of estimated market rental value. These differences in market rental value are paralleled by differences between participants and controls in the ability of program participants to achieve higher housing quality measured in terms of normative standards of physical housing quality and crowding. These latter differences, however, are sensitive to the definition of housing standards, making generalizations regarding increases by program participants in the ability to meet normative standards somewhat risky.

While the housing benefits provided by each program are significant, they are not generally the dominant benefits associated with program participation. Instead, increases in income available for goods and services other than housing comprise, for all programs other than Section 236, from roughly two-thirds to three-quarters of the economic value of tenant benefits.[18] These increases in income result from reductions in tenant rent burdens, which in turn result from program participation. In terms of household income, these income benefits represent an increase in income available for nonhousing goods and services of from roughly 6% to 11%. It is clear from the decomposition of benefits that most of the programs analyzed here are as much or more income transfer programs than housing programs.

In part because benefits are heavily weighted in terms of income, it is estimated that there are only modest consumer welfare losses associated with the package of benefits offered by each program. That is, the estimated cash equivalent value of program benefits is estimated to be close to the apparent market value of the package. This finding does not, however, suggest that all of these programs are efficient in delivering housing services to tenants. Even modest losses in consumer welfare resulting from inappropriate mixes of housing and income benefits may be cause for concern if there are not offsetting donor benefits accruing to providers of subsidies. While a reasonable presumption of donor benefits may be supposed to exist based on the apparent program impacts on tenants' ability to achieve normative housing standards, it remains to be seen whether the costs required to achieve them (consumer welfare losses and resource costs) are justifiable.

Aside from the estimated deadweight losses in consumer welfare that have been estimated here, two other costs require balancing against program benefits. The first, the loss in consumer sovereignty imposed by some programs, has been touched on in this review in terms of program impacts on tenant satisfaction. It was seen that levels of tenant satisfaction with neighborhoods were lower in the more geographically restrictive supply-oriented programs such as public housing and Section 236 than not only demand-oriented programs but also among control households. The second important cost is the resource cost of providing units in each program. While it has not been possible in this brief review to cover costs as well as benefits in housing programs, Mayo et al. (1980b: 71) indicate that at the time of the

Demand Experiment the annual costs of providing a new public housing or Section 236 unit were an average of 82% higher than the costs of providing a unit by way of housing allowances. Moreover, it was pointed out that economic trends affecting housing and construction markets apparent in the mid-1970s (and which have subsequently been continued) seemed likely to increase the cost differential further between supply-oriented and demand-oriented programs.

It is in light of these large cost differences between demand-oriented and supply-oriented programs that comparative benefits of housing programs must be evaluated. Thus, while this analysis points out some clear differences among programs in the benefits they provide, in the harsh light of large and pervasive program cost differences, differences in benefits provide only modest guidance in choosing an appropriate mix of housing policies. Based on the data analyzed as part of the Demand Experiment, it was apparent that a mix heavily weighted in favor of demand-oriented programs such as housing allowances, Section 23 Leased Existing Housing, or, by extension, Section 8 Existing Housing was clearly preferable to one weighted in favor of supply-oriented programs. Based on subsequent developments in housing and credit markets this conclusion remains valid in the early 1980s.

NOTES

1. These reports deal with (1) program outcomes in the areas of participation, housing consumption, location, and satisfaction; and (2) program costs, efficiency, and equity.

2. Data were collected on 722 households in Pittsburgh and 491 households in Phoenix that were residents of subsidized housing units in the three comparison programs during 1975. To sharpen comparisons, sampled populations were restricted in some ways. Public housing units samples were restricted to those in typical public housing subprograms—Conventional and Turnkey I programs, the two of which account for the bulk of public housing units provided during the late 1960s and early 1970s. Section 23 units were limited to those under which otherwise unsubsidized housing units from the existing privately owned housing stock were leased by public housing agencies. Section 236 units were limited to those in projects that contained at least some rent supplement units. For a complete description of the sample, the programs covered, and their characteristics, see Mayo et al. (1980a, Appendix II).

3. The payment formula for HGMS is $S = C^*(H) - bY$, where $C^*(H)$ is the payment standard C^* which is conditioned on household size (H) and on experimental site; Y is adjusted household income; and b is the benefit reduction ratio.

4. The possibility exists in any demand-oriented program that program participants may systematically underpay or, more likely, overpay relative to unsubsidized households in the private market.

5. For details of the calculation, see Mayo et al. (1980a: 61ff).

6. For example, fewer than 10% of subsidized units were estimated to be valued at more than the value of private housing in the seventy-fifth percentile of the rent distribution; most of that was Section 236 housing.

7. Subjective data provide strong confirmation of the degree to which benefits occur in the form of income rather than housing. In response to an open-ended question, "What are some of the things you like best about this housing program?" households in each comparison program generally responded overwhelmingly in terms of some aspect related to low program rents—"low," "less," "reasonable," or "cheaper" rent; "can afford to pay rent," "rent in line with income," or "utilities included in rent." Such responses occurred from 1.3 times for Section 236 to 8.8 times for Section 23 more often than those of the next most often cited program feature in Pittsburgh, and from 2.7 to 4.1 times more often in Phoenix. The relative frequency of responses to different program features closely paralleled estimated program impacts on income and housing.

8. Alternatively, Section 236 participants may be selected to a considerable degree from among households with above-average tastes for housing. Data did not permit examining such selection biases among participants in the various programs.

9. Of course, there may be no net loss to society as a whole if the donors of subsidized housing value the improvements in housing experienced by program recipients more than program recipients value their loss of consumer sovereignty.

10. Such a measure represents an approximation to the more theoretically appropriate Hicksian consumer's surplus, which was not calculated here because of data and estimation problems. In particular, when a Stone-Geary expenditure function was estimated that would, in theory, have permitted an exact Hicksian income equivalent to be calculated, parameters were such that the income equivalent was undefined for a substantial fraction of the population. This is a result of the functional form of the expression for the income equivalent under a Stone-Geary function and the inability to develop a satisfactory mechanism for distributing parameters across demographic groups. For more details, see Friedman and Weinberg (1980a: Appendix VII).

11. Estimates of housing changes presented here are smaller than those estimated by the National Housing Policy Review, thereby contributing to the smaller estimated deadweight losses. See U.S. Department of Housing and Urban Development (1973: 4-15ff).

12. For a review of housing demand literature, see Mayo (1981). In that review, price elasticities of housing demand of roughly −0.50 are commonly found, whereas Table 15.4 is based on an assumed price elasticity of −0.22.

13. See, for example, U.S. Department of Housing and Urban Development (1973), Kraft and Olsen (1977), and Murray (1975).

14. Actual administration of the Section 23 program varied significantly, however. In some areas (e.g., Pittsburgh), Section 23 was run as a program for large minority families and offered leased units only within a narrow geographical area. In Phoenix, sampled Section 23 units were drawn in effect from two different programs—an "original" and a "revised" Section 23, the former permitting participating households narrower choices than the latter (which, like its successor, Section 8, permits households to find their own units).

15. As in the case of physical standards, households must satisfy occupancy requirements to qualify initially for a subsidy. Continuous compliance is not, however, required to receive a subsidy.

16. For more general discussions of public housing in the postwar period, see deLeeuw and Jarotis (1970), Meehan (1975), and White, Merrill and Lane (1979).

17. Figures for public housing and Section 23 suggest nominal violations of the Brooke Amendment standard for some households at each site. This finding depends undoubtedly in part on rent and income measures used here as compared to such measures as defined by public housing agencies. For example, the Demand Experiment rent measure, which is used to define the analytical rent variable in all programs evaluated here, includes utilities. In Phoenix, most

public housing and Section 23 tenants either pay fuel utility bills directly to utility companies or have check-metering arrangments whereby tenants pay for excess usage. Such payments are included as rent by Phoenix PHAs. As a result, rent burdens calculated here are higher than program rent burdens.

18. In the two cities studied here, income benefits for Section 236 participants comprised from about 0%-40% of estimated economic benefits of program participation.

16

Lessons for Future Social Experiments

MARC BENDICK, Jr. and RAYMOND J. STRUYK

☐ IN ONE WAY at least, the Experimental Housing Allowance Program was an undertaking of major significance beyond the field of housing policy itself: It was a methodological landmark within the developing art and science of evaluating proposed public policies through formal social experiments. The federal government has invested in only a handful of national-scale social experiments, covering such issues as guaranteed incomes for the poor, national health insurance, "supported work" for the hard to employ, educational performance contracting, and peak load electricity pricing.[1] Thus, rarity alone makes each experiment noteworthy. But even in comparison with the other major social experiments, EHAP commands particular attention in terms of its size and complexity.

A major-scale social experiment requires investment of millions of taxpayers' dollars; the total budget for EHAP, for example, was $158 million. It also absorbed large amounts of a commodity even more scarce than financial resources: the political process' willingness to consider a major policy innovation. To utilize such resources with maximum efficiency, it is important that lessons be drawn from each experiment for future practices. This chapter presents what were, in our view, some important lessons from the EHAP experience for future social experimentation.[2]

Throughout its lifetime, EHAP was called on simultaneously to serve two quite different missions. The mission that initiated its existence and, at various times, rose to the fore was that of *program evaluation* of a specific approach to low-income housing assistance being considered for nationwide implementation. The other mission, which was largely conceived of by EHAP researchers themselves, was that of a vehicle for *basic research* on housing markets and housing consumers. The most general lesson of the EHAP experience is that the degree of compatibility between these two missions was less than might have been expected; in consequence, future social experiments might more effectively serve both missions if their design integrated the two approaches in more sophisticated ways.

Closely related to the tension between basic research and program evaluation is the tension between two research strategies for studying the effects of a social program in action. One strategy, which might be labeled that of *holistic experimentation*, involves simply implementing the program in its entirety in as close a resemblance to a full-scale implementation as possible and then observing its effects directly (netting out other influences through use of a control group). An alternative approach, which might be labeled *analytic experimentation*, uses the experimental approach only to estimate specific unknown parameters; the full impact of the program being experimented with is then estimated analytically, using these new empirical estimates at crucial steps in the analysis. We will argue that EHAP represented a major evolution from the former toward the latter strategy; and we will conclude that future social experiments will be made more useful by continuing to evolve in this direction.

ALTERNATIVES FOR STUDYING MARKET RESPONSES

We will first illustrate these points in the context of the Supply Experiment, EHAP's approach to studying the response of housing markets to an allowance program.

When EHAP was being designed, two alternative research strategies were considered for investigating market responses. One approach would have relied exclusively on computer simulations of urban housing markets, using models already being developed by The Urban Institute and the National Bureau of Economic Research.[3] With these models, the effects of housing allowances could be examined for a wide range of cities, both large and small, at relatively low cost (a few millions of dollars). The alternative approach was to field a full-scale operating allowance program that would generate a major demand stimulus in one or more housing markets and observe price and supply responses directly. This latter approach, which was implemented as the Supply Experiment, had the advantage of greater face validity than any purely analytical approach. However, it involved a very high cost—eventually $99 million—and even at that level of expenditure had to be restricted to only two housing markets, both of them relatively small.

Eventually, both research approaches were pursued; and we therefore are now in a position to see how accurately the cheaper alternative—computer simulation—matched the results of the vastly more expensive field operation. In particular, we can directly compare the results obtained from the Supply Experiment in its two operating sites—Green Bay, Wisconsin and South Bend, Indiana—with simulations of those same markets using The Urban Institute's Housing Market Model.[4]

At first glance, the predictions from the model and the field results from the Supply Experiment seem directly at odds: The model predicted a substantial rise in the price of low-income housing in South Bend (although none in Green Bay), while the actual field experience was that price rises occurred in neither site. However, examination of the basis of the model's predictions pinpointed two specific sources of the prediction error: unrealistic parameters for households' propensity to participate in the program and for households' propensity to spend incremental income on housing. When the model was rerun with those two household behavioral parameters reset to match those experienced in the field, the model's predictions accurately matched the observed reality.

The important point here is that the best estimates of the two key parameters in the model that required correcting could have come not from the Supply Experiment but from the Demand Experiment, which was specifically designed to investigate household behavior.[5] A combination of an analytical device—housing market simulation models—plus household information from the Demand Experiment could have provided the same answer provided by the Supply Experiment—at a cost of nearly $100 million.

Of course, the fielding of a large-scale program in Green Bay and South Bend did tend to give this answer considerably more credibility with decisionmakers than any mere "paper and pencil" analysis. On the other hand, there are severe limits to how far this point must be stressed. Budgetary considerations meant that only two sites could be included in the Supply Experiment, and both of them are medium-sized cities in the upper Middle West. The Supply Experiment provided no direct test of the market reactions to allowances in rural areas; or in the physically and sociologically different cities of the South and West; or, most crucially, in large, decaying urban centers such as Detroit or Newark, which are the focus of much public policy concern.[6] Extrapolation of Supply Experiment results to these diverse situations already requires an analytical step, such as a computer simulation. Furthermore, the Administrative Agency Experiment (and, to some extent, the Demand Experiment) already were providing at least limited field demonstration of the program in operation.

The two alternative research strategies thus might seem to be characterized as (a) answering a question with reasonable credibility (via computer simulation building on results from the Demand Experiment, plus field demonstrations in the Administrative Agency Experiment); or (b) answering the question with somewhat higher (but not universal) face validity, at an additional cost of about 20 times that of the analytic approach. With the benefit of a decade of hindsight since the original decisions were made—and the benefit of the EHAP experience itself—we would tend to favor the first option.

SEPARATING BASIC RESEARCH FROM EXPERIMENTAL PROGRAM EVALUATION

Although the primary purpose of the Supply Experiment was evaluation of housing allowances as a potential national program, it also simultaneously served as a major undertaking in basic research on housing markets. Thus, the data it generated and the analyses it produced must also be credited as returns from the $99 million investment. And there is little doubt that the analytical bounty to be produced for the vast data sets collected on Green Bay and South Bend will eventually be substantial. These files contain a more detailed, longitudinal picture of the supply side of housing markets than have ever been available before.

There are two drawbacks to the simultaneous pursuit of program evaluation and basic research objectives, however. The more minor of the two is that the data generated may not be ideally suited for basic research purposes. For example, from the point of view of non-EHAP research, the behavior of the markets may be contaminated by the allowance program. Likewise, nonrepresentative sets of observations emerge: Certain data were collected over time for only nonmoving households or for only participating households.

But the more major difficulty is the adverse effects on the program evaluation objective, particularly in terms of timeliness. The amount of data generated by the Supply Experiment was huge; it has been estimated to equal one-third the size of the nationwide 1970 census. Therefore, it should come as no surprise that lags of several years were experienced between the moment data were initially gathered in field surveys and when they became available for analysis. Even if data preparation had required virtually no elapsed time, EHAP would have faced difficulty turning out results fast enough to make them timely for public decision making.[7] Had data collection in the Supply Experiment been restricted solely to those directly required for the program evaluation objective, then the long lags encountered in data preparation would surely have been shortened, although obviously not eliminated.

Another cost of these delays was the excessive time for those monitoring the experiment to react to what should have been seen as early key patterns—especially lower than expected rates of participation and increases in rents. By the time the data were prepared and analyzed, there was little to do but continue following a data acquisition plan that was unlikely to provide a basis for further major policy findings. Had these facts been grasped sooner, monitoring efforts might have been shifted or substantial savings realized in the cost of the evaluation.[8]

The counterargument against such separation, of course, is that there are economies of scale in data collection that would be tragic to lose. The marginal costs of asking "just one more question" seem temptingly close to zero, when compared to the huge fixed costs of setting up the study in the first place. Moreover, the synergistic research benefits of having a broad range of information available in the same microdata set seem irresistibly attractive. But the Supply Experiment's experience of such long delays—and the fact that in virtually all parts of EHAP, vast proportions of the data gathered have not been touched for any research purpose—should tend to put some counterweight on the opposite considerations as well. *Dis*economies of scale in data management are a very real possibility; the costs of handling an individual piece of information can increase, rather than decrease, when the total mass of data reaches census-like proportions. And the true synergistic possibilities of a tremendously broad range of data must be estimated with some caution; the EHAP experience has yet to confirm that they are actually exploited to any substantial degree.[9]

In the future our preferred strategy for using the millions of dollars of resources needed to support another supply experiment would be to fund longitudinal panel studies.[10] For some goals of EHAP research, a national longitudinal panel of housing units—like the Annual Housing Survey but with the greater detail on landlords' income, and housing expenditures and investments of the Supply Experiment's four-year dwelling panel—would be the ideal vehicle. For other issues, longitudinal tracking of households, modeled on the Panel Study of Income Dynamics, or possibly an augmented version of it, would be more appropriate.

Such long-term data bases would obviously be fertile, continuing vehicles for basic research. They also would be the most appropriate sources of the basic behavioral estimates needed for the EHAP program evaluation role. To see why this is true, we must turn to a discussion of the Demand Experiment.

ALTERNATIVES FOR STUDYING HOUSEHOLD BEHAVIOR

The household behavioral phenomena the Demand Experiment was designed to examine—program participation, housing consumption, and residential mobility—are among the most complex behavioral phenomena social scientists might tackle. They involve such complexities as residential moves coinciding with changes in family composition; the discreteness and infrequency of changes in housing consumption; irregular patterns of savings and dissaving when families accumulate funds to finance changing from being renters to homeowners; and so forth. Such complexities make it difficult to get full and accurate revelation of a household's adjustment to an allowance program in an experiment of relatively short duration and in which the range of behavior by which the household could reveal its reactions were limited. In the Demand Experiment, a household could receive benefits for at most three years; likewise, renters could not become homeowners, and changes in family composition were not well tracked.[11]

The Supply Experiment, which allowed renters to become homeowners and which paid benefits for as much as five years, came closer to providing the opportunity to observe the behavior in its full complexity; but then it did not include a household control group. Moving even further in the same direction would involve switching to a research strategy of tracking households in longitudinal panels with no experimental stimulus, using natural variations across time and in the cross-section to generate sampling variation.[12] The adjustment of a longitudinal panel of households to their ongoing circumstances would more closely approximate the reactions of households to a permanent housing allowance program than would the short-term reaction of households to a temporary experimental stimulus.

Thus, in the Demand Experiment, as in the Supply Experiment, there is reason to believe that both the program evaluation objective and the basic research objective might well have been better served by devoting research resources to longitudinal panel studies of households, rather than to the more simpleminded direct experimentation with an allowance program per se.

Still, however, the need for a strong experimental component remains. A formidable amount of time would be required to develop the data in a longitudinal file and to have enough analytic experience with it to truly understand these complex dynamic processes. This is evident, for example, from the evolving appreciation of the dynamics of poverty coming from the work with the Panel Survey on Income Dynamics. Experimentation is, with its limitations, the shortest route to a rough answer.

Experimentation also must be relied on for analysis of those aspects of behavior that arise only when the program is implemented. One prominent example from EHAP comes readily to mind: the response of households whose dwellings failed the program's inspection for minimum quality and safety features. Three responses were possible: to fix up the unit (or attempt to get the landlord to do so), to move to a unit that would pass the inspection, or to cease trying to qualify for a subsidy. Longitudinal records would be largely silent on this issue.

CONCLUSIONS

None of the comments made in this chapter should be taken to denigrate the decisions made in 1970 in designing EHAP. Given the state of the art at that time, and given that no one knew what the results of the housing allowance experiments would be, the research strategy adopted at that time made considerable sense.[13] But today, with the benefit of 12 years of hindsight, we suggest that if the same research objectives were set forth today, serious consideration should be given to a *nonexperimental* approach. That approach would be more bold in relying explicitly on analytical interpretation of basic behavioral research (derived from longitudinal micro panels, assuming they existed) rather than on the more apparently direct one of experimentation. This analytical work would be kept quite separate from field demonstrations of the concept (on the model somewhere between the Administrative Agency and Demand experiments), which also would be a necessary part of developing a major policy initiative.

The challenge for policy analysts and researchers is to find political ways to "sell" to public decisionmakers the crucial, long-term investment in basic panel research. Is it realistic to think that the federal government might have been willing to spend $158 million on basic housing research rather than on an experiment linked closely to an imminent new national program? Probably not, and that is why researchers must learn to tie such panel studies closely to incipient policy issues, or as part of a package approach to more urgent issues. After all, the war on poverty of the late 1960s generated the Panel Study of Income Dynamics, as well as the negative income tax experiments; the need to evaluate the Comprehensive Employment and Training Act of the mid-1970s led to the rich National Longitudinal Studies of labor markets. Exercise of creativity in forging such linkages, and firm belief that the truly correct way to do program evaluation involves building from basic behavioral understanding, are essential to making this approach practical. But the potential payoffs from such efforts—for both basic research and program evaluation—demand no less.

NOTES

1. For an overview of these large-scale studies, see Ferber and Hirsch (1978). Boruch and Riechen (1975:142-145) list a sample of smaller experimental efforts in such fields as law enforcement, mental health, and social services.

2. This chapter discusses primarily the Demand Experiment and the Supply Experiment. We do not examine the Administrative Agency Experiment because it belongs primarily to the tradition of demonstration programs rather than that of social experiments. On the AAE and its controversial design, see Hamilton (1979), Kershaw and Williams (1981:302-319) and Struyk and Bendick (1981:41-44, 45-47).

3. These two models are described in Struyk and Bendick (1981:213-220).

4. The simulation results are reported in Vanski and Ozanne (1978) and Struyk and Bendick (1981:207-234).

5. In actually rerunning the model, values from the Supply Experiment were used to facilitate direct comparisons of outcomes. But the Demand Experiment was designed to provide estimates of such parameters.

6. On the representativeness of EHAP sites, see Struyk and Bendick (1981:329-338) and Lowry (1982b).

7. A housing allowance-type program—the Section 8 Lower-Income Rent Supplement Program for Existing Housing—was passed into law in 1974, when the only EHAP results available to be used in the program debates were early findings from the Administrative Agency Experiment. A similar pattern occurred with the New Jersey-Pennsylvania Negative Income Tax Experiment, when President Nixon's Family Assistance Plan came under active congressional consideration at a time when the only results available from the experiment were preliminary analyses of one year's worth of data. On the other hand, the Reagan administration has again returned to vouchers, partially on the basis of the evidence from EHAP.

8. For more on the notion of contingency designs in the research plans of long-term social experiments, allowing midcourse corrections of data-gathering strategies on the basis of early empirical findings, see Struyk and Bendick (1981:302-303).

9. The jury is still out on this subject, however. This is because the U.S. Department of Housing and Urban Development has gone to considerable lengths to make the entire EHAP data archives publicly available (through DUALABS, Inc.) and to advertise their availability and

encourage their use. Only the passage of time will tell if they receive extensive use (other than for reanalysis of the same questions addressed in the original analysis). The experience with the data sets from the Negative Income Tax Experiments, which have been available for several years, has not included very much creative, broad-ranging use.

10. It is probably realistic to think of reprogramming to alternative research uses only the costs of EHAP exclusive of payments to households. This amount totals $103 million for EHAP overall and $59 million for the Supply Experiment alone.

11. Watts (1981:esp. 53-57) discusses in detail the extent to which the EHAP experiments were less sophisticated than was desirable in modeling these aspects of the housing adjustment process.

12. In fact, within EHAP, considerable research on household reactions to allowances was conducted on the cross-sectional sample of households in the Demand Experiment's baseline (preexperimental) survey. See references H-22 through H-31 in Struyk and Bendick (1981: 395-396).

13. This is the same summary judgment stated at the end of our more detailed review of the EHAP design presented as Chapter 2 of Struyk and Bendick (1981).

17

Implications for Housing Policy

EDGAR O. OLSEN

☐ DESPITE THE HUGE VOLUME of research under the Experimental Housing Allowance Program, its major implications for housing policy can be stated briefly. First, we should seriously reconsider the desirability of providing housing subsidies in any form. Second, if we decide to provide housing subsidies, we should replace all current programs involving the construction of new units or the rehabilitation of units selected by bureaucrats with some type of housing allowance program. This chapter develops these two implications.

HOUSING SUBSIDIES OR UNCONDITIONAL CASH GRANTS?

The first and most fundamental implication of EHAP for housing policy is that we should reconsider the desirability of providing housing subsidies in any form.

Let us define an in-kind subsidy to be one which changes consumption patterns in a way different from a program of cash grants with no strings attached.[1] Presumably, an in-kind subsidy is not a housing subsidy unless it induces its recipient to consume more housing services generally or to occupy housing that is better in certain respects than the recipient would choose if given an equally costly unconditional cash grant. It is a well-known proposition in economics that a recipient will prefer an unconditional cash grant to any in-kind subsidy with the same cost to the giver because it enables the recipient to choose the consumption bundle that would be chosen with the in-kind subsidy but makes available at least one preferred alternative.[2]

Since recipients prefer an unconditional cash grant to a housing subsidy, it is difficult to justify housing subsidies unless taxpayers receive tangible benefits particularly related to the housing of recipients or care about recipients but think that these people undervalue housing. The existence of tangible benefits is frequently asserted. For example, it has been argued that better housing for low-income families leads to better health for its occupants and, since some diseases are contagious, to better health for the middle- and upper-income families with whom they come in contact. Available evidence suggests that some such externalities exist but that their magnitudes are small (Weicher, 1979:489-492).[3] If the goal of housing subsidies is to make both recipients and taxpayers better off, it is doubtful that substantial expenditures can be justified on the basis of these externalities alone. I believe that the major rationale for housing subsidies is the existence of paternalistic altruists, but it must be admitted that the extent and depth of this sentiment have never been seriously studied.

The contribution of EHAP to the discussion of the desirability of housing subsidies is to provide evidence suggesting that supporters of these subsidies have a distorted perception of the housing conditions of low-income households and of the reasons families are poorly housed. The major type of housing allowance tested offered each eligible family a cash grant on the condition that it occupy housing meeting certain standards. In deciding which standards to use, a search was made for scientific evidence concerning the effects of specific housing defects on the health, safety, and behavior of occupants. Municipal housing codes in the experimental sites and similar localities were reviewed and compared with the model codes promulgated by various organizations (Lowry et al., 1974:13-15). Neither existing nor model codes are based on substantial scientific evidence, and the evidence available does not suggest an obvious choice of standards. The Demand Experiment adopted a modified version of the American Public Health Association-Public Health Service Recommended Housing Maintenance and Occupancy Ordinance (Bakeman et al., 1980). In the Supply Experiment, the Building Official's Code of America, with minor modifications, was used. It seems reasonable to presume that model housing codes are designed by people who have particularly strong feelings concerning the value of better housing for low-income families. To the extent that such families already occupy housing meeting these standards, the argument for housing subsidies based on taxpayer preferences is undermined.

It has been estimated that about half of the eligible families in Supply Experiment sites occupied housing meeting its standards and about a fourth of the eligible families in the Demand Experiment sites occupied housing meeting its more stringent standards prior to receiving assistance (U.S. Department of Housing and Urban Development, 1980:107).[4] Even at the lowest income levels, many households occupied units meeting the standards of the Demand Experiment (Budding, 1980:31, A-100). I draw two conclusions from these findings. First, almost all low-income families are *able* to occupy housing meeting the standards embodied in model housing codes. Some choose not to do so. Second, low-income families are not as poorly housed as is widely believed.[5] If these conclusions are accepted, it seems reasonable to believe that many taxpayers who favor housing subsidies do so based on erroneous assumptions, and so a reconsideration of the desirability of housing subsidies versus unconditional cash grants is in order.

DESIRABLE PROPERTIES OF A HOUSING PROGRAM

If, in the light of better information, many taxpayers still believe they receive tangible benefits especially related to the housing of low-income families or care about these families and think they undervalue housing, then housing subsidies are necessary to achieve an efficient allocation of resources preferred by everyone to the allocation in the absence of government action, provided that the conditions governing these subsidies can be enforced cheaply enough (Olsen, 1981a:159-165). In this case it is important to consider what properties a housing program must have in order to be consistent with taxpayer preferences.[6]

Six properties seem particularly important. First, the program must induce the worst-housed families at each income level to occupy better housing than they would choose if they were given equally costly cash grants with no strings attached. Many low-income families occupy housing which meets the standards that have been used in housing programs. Typically this is achieved by spending a large proportion of income on housing or, more precisely, on those attributes of housing covered by the standards. Presumably, paternalistic altruists feel that families with the highest rent-income ratio overvalue housing relative to other goods and want to distort their consumption patterns away from housing. Second, families that are the same with respect to characteristics of interest to taxpayers should be offered the same assistance. Third, the greatest assistance should go to the neediest families. Among families that are the same in other respects, income including transfers and the net rental value of consumer durables is perhaps the best available indicator of need, and so assistance should be greater for families with lower incomes.[7] Fourth, the subsidy should be zero at the upper income limit for eligibility. If it is positive, then subsidized families with incomes below the limit will be able to consume more than some ineligible families. This is inequitable and reduces the incentive to earn for families with preprogram incomes somewhat above the limit. Fifth, the subsidy should decline by much less than a dollar for each additional dollar earned. Since most low-income families participate in several income-conditioned subsidy programs, they already face substantial marginal tax rates (Browning, 1976). To make these families eligible for another income-conditioned subsidy with a substantial subsidy-reduction rate would virtually eliminate their incentive to earn.[8] Finally, the housing services provided to participants should be produced at the lowest possible cost.

Other goals for housing programs have been suggested: for example, reducing racial segregation in housing or stabilizing new construction. The rationales for achieving these goals are different from those underlying housing subsidies to low-income families, and I believe they are best achieved by other means.

HOUSING ALLOWANCES OR NEW CONSTRUCTION PROGRAMS?

The central issue in housing policy is the relative merits of programs involving the construction of new units or the rehabilitation of units selected by bureaucrats on the one hand and those involving assistance to families who are free to choose any unit meeting housing standards on the other. The effects of programs within each category are similar, and the differences between programs in different categories are marked. Since the rehabilitation programs account for such a small share of HUD's budget and have been subject to much less research than the other types of programs, they will be ignored in the remainder of this chapter.

EHAP is the major source of information about the effects of a housing allowance program but accounts for only a small fraction of the high-quality research on the effects of current programs.[9] Without the non-EHAP information, considerable caution would be warranted in judging the merits of a housing allowance program compared with existing programs. With it we can be reasonably confident in our judgment. This section summarizes what is known about the extent to which housing allowance and new construction programs are consistent with plausible assumptions about taxpayer preferences.

EFFECTIVENESS

The scarce evidence indicates that a substantial majority of participants in the public housing program occupy housing better than they would choose if they were given cash grants with no strings attached (Kraft and Olsen, 1977:58-60; Olsen and Barton, 1982:28-29).[10] Although evidence is not available for other new construction programs, this is also likely to be true for them because the average age of units under these programs is much less than under the public housing program. What is not known is whether the families served by these programs were the worst-housed families at each income level. Therefore, we cannot say whether the new construction programs have had the desired effects in this regard.

As already noted, the major type of housing allowance tested in the experiment was the housing gap with minimum standards. The effects of such a housing allowance program depend in part on the magnitude of the grant and the standards used (Cronin, 1981:89-101; 1981a:129-157). In the Supply Experiment, about half of the participants occupied housing meeting the standards prior to receiving a subsidy, and so it is reasonable to conclude that this particular housing allowance program was equivalent to an unconditional cash grant for these families. About two-thirds of the participants in the Demand Experiment lived in housing meeting its standards prior to the program (Kennedy and MacMillan, 1980:129; see also chap. 6). Evidence suggests that few of the other families would have met the standards in response to an equally costly unconditional cash grant (Friedman and Weinberg, 1980:229; see also chap. 8). I conclude that a minimum standards program similar to those tested in EHAP will be equivalent to an unconditional cash grant for a substantial proportion of participants, but it will induce many others to occupy housing that is better in terms of the characteristics covered by the standards than they would choose in response to unconstrained cash grants.

It has been found in EHAP that among similarly situated families those with the worst housing in terms of the program's standards have the lowest participation rates (Goedert, 1979:98-99). However, since no evidence is available on the performance of current housing programs in this regard, this gives us no reason to prefer these programs to the minimum standards housing allowance programs tested. Furthermore, this shortcoming can be overcome without increasing the cost of the program by using another type of housing subsidy that does not involve the construction of new units. Each eligible family could be offered a subsidy consisting of a fixed payment G independent of housing expenditure plus a fraction K of housing expenditure up to some maximum amount. For housing expenditure beyond this point, the subsidy would decrease by some amount C for each additional dollar spent on housing. To avoid a notch, G, K, and C should be smaller for households with higher incomes and zero at the upper income limit for eligibility. Under this program, the worst-housed families with participation costs less than the fixed payment G would participate. For these families, the program provides a special inducement to improve their housing by lowering its price to them. For the best-housed eligible families, it provides a special inducement to increase their consumption of other goods (that is, to decrease their rent burden).[11]

EQUITY

The current system of housing programs in the United States is fraught with inequities.[12] Among eligible households that are willing to participate and that have the same characteristics, a fortunate minority receive large subsidies while the majority receive nothing. Among similar households that are served, the variance in the subsidy is large primarily because there is a great variance in the desirability of different subsidized units but little variance in the rent paid by tenants. Households that have incomes near the upper limits for eligibility and occupy the best subsidized units receive substantial subsidies, with the result that their effective incomes are greater than those of the poorest ineligible households.

By contrast, a universal housing allowance program offers the same subsidy to all families that are the same in all respects judged to be relevant for housing policy. The subsidy is greatest for households with the lowest income and greatest size. Since the upper income limit for eligibility for families of each type is the income at which the subsidy is zero, a notch is avoided.[13]

EFFICIENCY

One of the most important findings of EHAP is that it costs substantially more to provide housing under HUD's two largest new construction programs, public housing and Section 236, than to provide equally desirable housing in the private market (Mayo et al., 1980b). If these were the only estimates in existence, considerable caution would be justified in acting on them. However, all previous estimates for these programs (e.g., U.S. Department of Housing and Urban Development, 1974: chap. 4; Olsen and Barton, 1982:25) and a subsequent estimate for HUD's fastest growing new construction program Section 8 (Weinberg, 1982:Figure 1), indicate their excessive cost.

Many explanations of this finding have been suggested. Some involve program features that provide incentives for inefficiency. For example, until 1970 the federal government paid the development costs of public housing projects while tenants and local taxpayers paid the operating costs. It is reasonable to believe that local housing authorities are sensitive to costs borne by local residents but insensitive to costs borne by other taxpayers. If so, local authorities would build housing as maintenance-free as possible, even though the extra development cost is much greater than the present value of the savings in maintenance costs. In 1970 Congress authorized HUD to reimburse housing authorities for the excess of their operating costs over their revenues. This reduced the price of maintenance and rehabilitation faced by housing authorities to zero. Much to the surprise of Congress and HUD, the operating costs of these authorities shot upward. By 1976 federal operating subsidies amounted to about 40% of the operating costs of public housing units, even though Congress had placed a cap on federal operating subsidies. Since maintenance and rehabilitation were free to local authorities during the early 1970s, it is reasonable to believe they did much that was not worth its cost.

Another explanation of the estimated inefficiency of the new construction programs involves the state of the housing market at the time the subsidized units are built. If comparable unsubsidized units are not being built at this time, it is because they cannot compete with existing units. That is, their costs, including a competitive rate of return, cannot be recouped at existing and expected future rent levels. This indicates that the vacancy rate among units of this overall quality is high, and therefore public construction makes no more sense than private construction.

One explanation of the inefficiency is common to all public production of goods and services. If a businessman finds a way to produce the same output at a lower cost, he gets to keep the difference. If an administrator of a housing program does the same, he does not get to keep the difference. Similar statements can be made about producing a better product at the same price. The effect of this difference in incentives between private and public providers of housing is obvious.

A housing allowance program avoids the inefficiencies of the new construction programs by subsidizing households and not requiring them to occupy newly built units.

MAJOR OBJECTIONS REBUTTED

The major objections to an entitlement housing allowance program have been that it would result in significantly higher rents without leading to much better housing and that it would be too costly.

The argument that landlords of allowance recipients would raise rents without upgrading their units is easily dismissed. In the Supply Experiment, recipients who remained in preprogram housing that already satisfied program standards experienced virtually no change in their rents. Even nonmovers whose units had to be improved in order to participate experienced small rent increases (Rand Corporation, 1978:82; see also chap. 10).

Few experts on the workings of housing markets believed that the effects of an allowance program on rents would be limited to recipients. Instead, most believed that the rents of similar units would be affected similarly. The rents of the least desirable units would fall and rents of modest units meeting the standards would rise at least in the short run. Disagreements among experts concerned mainly the magnitude of these short-run price changes (e.g., President's Committee on Urban Housing, 1968:71-72; Olsen, 1969:619-621; deLeeuw and Ekanem, 1973).

Evidence from the Supply Experiment shows beyond reasonable doubt that an entitlement housing allowance program similar to the one tested will have no significant effects on rents even in the short run (Barnett and Lowry, 1979; Mills and Sullivan, 1981; see also chap. 11). Reasons for the program's small effect on rents are easy to find. Eligible families account for only a small fraction of the demand for housing services in a given housing market, and many of these families choose not to participate. So even a large increase in demand by participants will have a small effect on aggregate demand for housing services, and this small increase occurs gradually over time because not all families respond instantly to the offer of a subsidy. Furthermore, it is clear that even over short periods suppliers are willing to make many changes in existing units in response to small changes in the profitability of housing with different characteristics.

Those who argue that an entitlement housing allowance program would be too costly fail to appreciate its flexibility. By adjusting the subsidy schedule or the standards or both, an entitlement program can be designed to have *any* cost desired. For example, it could be designed to have a cost less than the current system of housing programs.

CONCLUSIONS

Evidence from the Experimental Housing Allowance Program and studies of current housing programs support two propositions about housing policy. If we want to help low-income families, think that these families know better than others what is good for them, and receive no tangible benefits from their improved housing, then current housing programs should be terminated. The money currently devoted to housing subsidies should be used to expand cash assistance or to provide relief to taxpayers or both. If we want to help low-income families and have a special interest in inducing them to occupy better housing than they would choose in response to equally costly cash grants, then current housing programs should be replaced with an entitlement housing allowance program. Such a program avoids the inefficiencies of the new construction programs by subsidizing households and not requiring them to occupy newly built units. It reduces the inequities of these programs substantially by offering equally situated households the same grant under the same conditions and providing a subsidy to all eligible families that wish to participate.

NOTES

1. If we were to define an in-kind subsidy as a subsidy that changes the recipient's *alternatives* in a way that is different from the way they are changed by an unconditional cash grant, the two types of subsidies could have the same effect on consumption patterns. For example, if we offer a family a cash grant of G dollars on the condition that it consumes at least H units of a particular good, the family may choose a consumption bundle involving more than H of this good. In this case it will be consuming the same combination of goods as it would choose with an unconditional cash grant because the quantity constraint is not binding. The food stamp program (since the elimination of the purchase requirement) is a good example of such a subsidy.

2. Several caveats should be mentioned in connection with this theorem. First, the usual proof involves two implicit assumptions. It is assumed that the administrative costs are the same for the two programs. In fact, an in-kind subsidy is more expensive to administer because it has an additional administrative function—namely, enforcing the conditions. It is also assumed that if the in-kind subsidy involves public production of the good as in public housing, the public and

private producers are equally efficient. Available evidence strongly suggests that housing produced under government programs involving the construction of new units is more costly than equally desirable housing produced in the unsubsidized private sector (e.g., Mayo et al., 1980b; Olsen and Barton, 1982:25). These considerations reinforce the conclusion of the theorem. The second caveat works in the opposite direction. No existing or proposed program of cash grants provides unconditional grants. The size of the grant always depends on the amount the family earns. This type of cash grant is itself an in-kind subsidy, and the recipient may prefer some other type of in-kind subsidy with the same cost to the giver (Corlett and Hague, 1953-1954). To the best of my knowledge, no one has argued for housing subsidies on these grounds.

3. The majority of studies that are considered to be part of the literature on the social costs of slums do not deal with externalities at all. They estimate, for example, the relationship between a person's housing and his own health. The absence of a relationship proves the absence of an externality along these lines, but the presence of a relationship does not show that an externality exists. Many of these studies are also of extremely low quality, in that they do not account for nonhousing determinants of the variable to be explained.

4. Applying the two sets of standards to a sample of units from the four sites revealed that the Demand Experiment's standards produced a failure rate 22.5 percentage points higher than the Supply Experiment's standards (Valenza, 1977:18).

5. Budding seems to reach the opposite conclusion, but the disagreement may be more apparent than real. He is referring to the beliefs of people who are familiar with previous estimates of the extent of "housing inadequacy," while I refer to the beliefs of all supporters of housing subsidies. I doubt that many taxpayers are aware of the previous estimates. I do, however, have many objections to Budding's analysis—for example, his failure to recognize that housing deprivation is not an objective concept and his inclusion of a high rent-to-income ratio as a housing defect.

6. See Olsen (1982:99-102) for an elaboration.

7. Income is not an ideal indicator because it depends partly on effort. We would like to know what people could earn with equal effort, but this is difficult to measure.

8. The typical subsidy-reduction rate in existing housing programs is, and in EHAP was, 25%. It could be argued that this rate could not be lowered without increasing the cost of these programs to taxpayers. Such is not the case. The subsidy available to families with no income could be lowered so that the cost to taxpayers is unchanged. It has been argued that this change would prevent families from occupying units meeting program standards, but recent evidence (Olsen and Reeder, 1981) indicates that this is not correct, even for the program that provides the smallest subsidy to households with no income, the Section 8 Existing Housing Program.

9. See, for example, De Salvo (1975), Kraft and Olsen (1977), Murray (1975), Olsen and Barton (1982), U.S. Department of Housing and Urban Development (1974), and Weinberg (1982).

10. An important defect of almost all studies of the effects of housing programs is that they fail to make this comparison. Instead, consumption patterns in the presence of the program are compared with those in its absence. The limited relevance of this comparison is easy to see. Suppose the government offered to pay certain families a fixed fraction of their expenditures on all goods except good X and that the price elasticity of demand for non-X is less than 1. Consumption of good X would increase, and the percentage increase in consumption of good X could exceed that for other goods. Nevertheless, this is not a successful program for subsidizing good X because it results in less consumption of this good than would an equally costly unrestricted cash grant.

11. See Olsen (1981b:369-373) for a discussion of the rationales for and objections to this proposal.

12. Olsen (1982:104-105) and Olsen and Barton (1982:29-34) summarize the evidence underlying this subsection. Although EHAP has provided little information relevant to the discus-

sion of equity, no consideration of the merits of a housing allowance program compared with the current programs is complete without it.

13. The limited enrollment housing allowance program proposed by the Reagan administration for fiscal year 1982-1983 will have little immediate effect on the overall equity of the system of housing programs. Most of the families served would be current participants in other HUD programs, but only a small fraction of the participants in current programs would receive housing allowances in place of their previous subsidy. The majority will be from the Section 8 Existing Housing program, which is the most equitable of the current programs (Bureau of National Affairs, 1982:722-723). The allowance program does promote equity by excluding from eligibility the richest families eligible for the current programs, but it will serve only a small fraction of eligible families that are not served by other programs.

References

AARON, H. J. (1981) "Policy implications: a progress report," pp. 67-112 in K. L. Bradbury and A. Downs (eds.) Do Housing Allowances Work? Washington, DC: Brookings Institution.

Abt Associates Inc. (1976) Third Annual Report of the Administrative Agency Experiment. Cambridge, MA: Author.

______ (1973) Experimental Design and Analysis of the Demand Experiment. Cambridge, MA: Author.

ALLEN, G. E., J. J. FITTS, and E. S. GLATT (1981) "The Experimental Housing Allowance Program," in K. L. Bradbury and A. Downs (eds.) Do Housing Allowances Work? Washington, DC: Brookings Institution.

ATKINSON, R., W. L. HAMILTON, and D. MYERS (1980) Economic and Racial/Ethnic Concentration in the Housing Allowance Demand Experiment. Cambridge, MA: Abt Associates Inc.

BAKEMAN, H., C. A. DALTO, and C. S. WHITE, Jr. (1980) Minimum Standards Requirements in the Housing Allowance Demand Experiment. Cambridge, MA: Abt Associates Inc.

BARNETT, C. L. (1979) Using Hedonic Indices to Measure Housing Quality. Report R-2450-HUD, October. Santa Monica, CA: Rand Corporation.

______ and I. S. LOWRY (1979) How Housing Allowances Affect Housing Prices. Report R-2452-HUD, September. Santa Monica, CA: Rand Corporation.

BARNETT, C. L. and C. W. NOLAND (1981) The Demand for Housing Space and Quality. Working Draft WD-630-HUD, July. Santa Monica, CA: Rand Corporation.

BIRCH, D. L. et al. (1973) America's Housing Needs: 1970-1980. Cambridge, MA: Joint Center for Urban Studies of MIT and Harvard University.

BORUCH, R. F. and H. W. RIECHEN [eds.] (1975) Experimental Testing of Public Policy. Boulder, CO: Westview Press.

BROWN, L. A. and E. G. MOORE (1970) "The intra-urban migration process: a perspective." Geografiska Annaler 52B.

BROWNING, E. K. (1976) "How much equality can we afford?" Public Interest 43 (Spring): 90-110.

BUDDING, D. W. (1980) Housing Deprivation Among Enrollees. Cambridge, MA: Abt Associates Inc.

______ (1977) Inspection: Implementing Housing Quality Requirements in the Administrative Agency Experiment. Cambridge, MA: Abt Associates Inc.

Bureau of National Affairs (1982) Housing and Development Reporter, February 15.

BURNS, L. and L. GREBLER (1977) The Housing of Nations. New York: Macmillan.

CARTER, G. M. and S. L. BALCH (1981) Measuring Eligibility and Participation in the Housing Assistance Supply Experiment. Report R-2780-HUD, September. Santa Monica, CA: Rand Corporation.

CARTER, G. M. and J. C. WENDT (1982) Participation in an Open Enrollment Housing Allowance Program: Evidence from the Housing Assistance Supply Experiment. Report R-2783-HUD. Santa Monica, CA: Rand Corporation.

COLEMAN, S. B. (1982) How Housing Evaluations Affect Participation in a Housing Allowance Program. Report R-2781-HUD. Santa Monica, CA: Rand Corporation.

Congressional Budget Office (1979) The Long Term Costs of Lower Income Housing Assistance Programs. Washington, DC: Government Printing Office.

______ (1978) Federal Housing Policy: Current Programs and Recurring Issues. Washington, DC: Government Printing Office.

CONLISK, J. and H. WATTS (1969) "A model for optimizing experimental design for estimating response surfaces." Proceedings of the American Statistical Association, Social Sciences Section, pp. 150-159.

CORLETT, W. J. and D. C. HAGUE (1953-1954) "Complementarity and the excess burden of taxation" Review of Economic Studies 21:21-30.

CRONIN, F. J. (1981a) "Consumption responses to constrained programs," in R. J. Struyk and M. Bendick (eds.) Housing Vouchers for the Poor. Washington, DC: Urban Institute.

______ (1981b) "Participation in the Experimental Housing Allowance Program," in R. J. Struyk and M. Bendick (eds.) Housing Vouchers for the Poor. Washington, DC: Urban Institute.

DeLEEUW, F. and H. F. EKANEM (1973) "Time lags in the rental housing market." Urban Studies 10 (February): 39-68.

______ (1971) "The supply of rental housing." American Economic Review 61, 5:806-817.

DeLEEUW, F. and E. L. JAROTIS (1970) Operating Costs in Public Housing: A Financial Crisis. Washington, DC: Urban Institute.

DeLEEUW, F. and R. J. STRUYK (1975) The Web of Urban Housing. Washington, DC: Urban Institute.

DESALVO, J. S. (1975) "Benefits and costs of New York City's middle-income housing program." Journal of Political Economy 83 (August): 791-805.

DICKSON, D. E. (1977) Certification: Determining Eligibility and Setting Payments Levels in the Administrative Agency Experiment. Cambridge, MA: Abt Associates Inc.

DOLBEARE, C. N. (1974) "The housing stalemate." Dissent, Fall.

DOWNS, A. (1973) Federal Housing Subsidies: How Are They Working? Lexington, MA: D.C. Heath.

ELLICKSON, P. L. (1981) Who Applies for Housing Allowances? Early Lessons from the Housing Assistance Supply Experiment. Report R-2632-HUD, August. Santa Monica, CA: Rand Corporation.

______ (1978) Public Knowledge and Evaluation of Housing Allowances: St. Joseph County, Indiana, 1975. Report R-2190-HUD, February. Santa Monica, CA: Rand Corporation.

______ and D. E. KANOUSE (1979) Public Perceptions of Housing Allowances. Report R-2259-HUD, September. Santa Monica, CA: Rand Corporation.

FARLEY, R., S. BIANCHI, and D. COLASANTO (1979) "Barriers to the racial integration of neighborhoods: the Detroit case." Annals of the American Academy of Political and Social Science 441:97-113.

FARLEY, R., H. SHUMAN, S. BIANCHI, D. COLASANTO, and S. HATCHETT (1978) "Chocolate city, vanilla suburbs: will the trend toward racially separate communities continue?" Social Science Research 7:319-344.

FEINS, J. D., R. G. BRATT, and R. HOLLISTER (1981) Final Report of a Study of Racial Discrimination in the Boston Housing Market. Cambridge, MA: Abt Associates Inc.

FERBER, R. and W. Z. HIRSCH (1978) "Social experimentation and economic policy: a survey." Journal of Economic Literature 16 (December):1379-1414.

FOLLAIN, J. R. and S. MALPEZZI (1980) Dissecting Housing Value and Rent: Estimates of Hedonic Indices for Thirty-Nine Large SMSAs. Working Paper 249-17, February. Washington, DC: Urban Institute.

FRIEDEN, B. S. (1980) "What have we learned from the Housing Allowance Experiment?" Habitat International 5:227-254.

FRIEDMAN, J. and D. H. WEINBERG (1982) "Housing consumption under a constrained income transfer." Journal of Urban Economics 11:253-271.

——— (1981) "The demand for rental housing: evidence from the Housing Allowance Demand Experiment." Journal of Urban Economics 9:311-331.

——— (1980a) The Demand for Rental Housing: Evidence from a Percent of Rent Housing Allowance. Cambridge, MA: Abt Associates Inc.

——— (1980b) Housing Consumption Under a Constrained Income Transfer: Evidence from a Housing Gap Housing Allowance. Cambridge, MA: Abt Associates Inc.

——— (1980c) "Housing consumption in an experimental housing allowance program: issues of self-selection and housing requirements," in E. Stromsdorfer and G. Farkas (eds.) Evaluation Studies Review Annual 5. Beverly Hills, CA: Sage.

GOEDERT, J. E. (1979) "Earmarking housing allowances: the trade-off between consumption and program participation." Working Paper 249-19, May. Washington, DC: Urban Institute.

HAMILTON, W. L. (1979) A Social Experiment in Program Administration: The Housing Allowance Administrative Agency Experiment. Cambridge, MA: Abt Books.

HAMILTON, W. L., D. W. BUDDING, and W. L. HOLSHOUSER, Jr. (1977a) Administrative Costs of Alternate Procedures: A Compendium of Analysis of Direct Costs in the Administrative Agency Experiment. Cambridge, MA: Abt Associates Inc.

——— (1977b) Administrative Procedures in a Housing Allowance Program: The Administrative Agency Experiment. Cambridge,MA: Abt Associates Inc.

HANSEN, E. W., C. D'ARC, P. BOREN, and A. W. WANG (1982) User's Guide to HASE Data: Vol. I, Overview; Vol. II, The Survey Files; Vol. III, The Program Files. Report R-2692-HUD, April. Santa Monica, CA: Rand Corporation.

HARTMAN, C. and D. KEATING (1974) "The housing allowance delusion." Social Policy 4 (January-February).

HASE Staff (1981) Proceedings of the General Design Review of the Housing Assistance Supply Experiment. Note N-1035-HUD, January. Santa Monica, CA: Rand Corporation.

——— (1980) Supplemental Design Papers for the Housing Assistance Supply Experiment. Note N-1037-HUD, October. Santa Monica, CA: Rand Corporation.

HAUSMAN, J. A. and D. A. WISE (1981) "Technical problems in social experimentation: cost versus ease of analysis." Presented at the National Bureau of Economic Research Conference on Social Experimentation, March.

HELBERS, L. and J. L. McDOWELL (1981) Determinants of Housing Repair and Improvement. Report R-2777-HUD. Santa Monica, CA: Rand Corporation.

HEILBRUN, J. (1981) Urban Economics and Public Policy. New York: St. Martin's.

HILLESTAD, C. E. and J. L. McDOWELL (1982) Measuring Neighborhood Change Due to Housing Allowances. Report R-2302-HUD. Santa Monica, CA: Rand Corporation.

HOLSHOUSER, W. L., Jr. (1977) Supportive Services in the Administrative Agency Experiment. Cambridge, MA: Abt Associates Inc.

——— (1976) Report on Selected Aspects of the Jacksonville Housing Allowance Experiment. Cambridge, MA: Abt Associates Inc.

KAIN, J. F. and W. C. APGAR, Jr. (1977) Simulation of the Market Effects of Housing Allowances, Vol. II: Baseline and Policy Simulation for Pittsburgh and Chicago. New York: National Bureau of Economic Research.

KAIN, J. F. and J. M. QUIGLEY (1975) Housing Markets and Racial Discrimination. New York: National Bureau of Economic Research.

KATAGIRI, I. and G. T. KINGSLEY [eds.] (1980) The Housing Allowance Office Handbook. Note N-1491-HUD. Santa Monica, CA: Rand Corporation.

KENNEDY, S. D. (1980) The Final Report of the Housing Allowance Demand Experiment. Cambridge, MA: Abt Associates Inc.

______ and J. MacMILLAN (1980) Participation Under Alternative Housing Allowance Programs: Evidence from the Housing Allowance Demand Experiment. Cambridge, MA: Abt Associates Inc.

KENNEDY, S. D. and S. R. MERRILL (1979) "The use of hedonic indices to distinguish changes in housing and housing expenditures: evidence from the Housing Allowance Demand Experiment." Presented at the Research Conference on the Housing Choices of Low-Income Families, Washington, D.C., March.

KENNEDY, S. D., T. K. KUMAR, and G. WEISBROD (1977) Participation Under a Housing Gap Form of Housing Allowance. Cambridge, MA: Abt Associates Inc.

KERSHAW, D. and R. WILLIAMS (1981) "Administrative lessons," pp. 285-338 in K. L. Bradbury and A. Downs (eds.) Do Housing Allowances Work? Washington, DC: Brookings Institution.

KING, A. T. (1976) "The demand for housing: integrating the roles of journey-to-work, neighborhood quality, and prices," in N. E. Terlekyz (ed.) Household Production and Consumption. New York: Columbia University Press.

______ and P. MIESZKOWSKI (1973) "Racial discrimination, segregation, and the price of housing." Journal of Political Economy 81 (May-June):590-606.

KINGSLEY, G. T. and P. SCHLEGEL (1982) Housing Allowances and Administrative Efficiency. Note N-1741-HUD. Santa Monica, CA: Rand Corporation.

KINGSLEY, G. T., S. N. KIRBY, and W. E. RIZOR (1982) Administering a Housing Allowance Program: Findings from the Housing Assistance Supply Experiment. Note N-1846-HUD. Santa Monica, CA: Rand Corporation.

KRAFT, J. and E. O. OLSEN (1977) "The distribution of benefits from public housing," in F. T. Juster (ed.) The Distribution of Economic Well-Being. New York: National Bureau of Economic Research.

LINDSAY, D. S. and I. S. LOWRY (1980) Rent Inflation in St. Joseph County, Indiana. Note N-1468-HUD, November. Santa Monica, CA: Rand Corporation

LOWRY, I. S. [ed.] (1983) Experimenting with Housing Allowances. Cambridge, MA: Oelgeschlager, Gunn & Hain.

______ (1982a) Experimenting with Housing Allowances: Executive Summary. Report R-2880-HUD, April. Santa Monica, CA: Rand Corporation.

______ (1982b) Rental Housing in the 1970s: Searching for the Crisis. Note N-1832-HUD, January. Santa Monica, CA: Rand Corporation.

______ (1980a) Contingency Planning for the Supply Experiment. Note N-1036-HUD, October. Santa Monica, CA: Rand Corporation.

______ [ed.] (1980b) The Design of the Housing Assistance Supply Experiment. Report R-2630-HUD, June. Santa Monica, CA: Rand Corporation.

______ [ed.] (1980c) General Design Report, Supplement. Note N-1052-HUD, December. Santa Monica, CA: Rand Corporation.

______ (1980d) Preliminary Design for the Housing Assistance Supply Experiment. Note N-1027-HUD, July. Santa Monica, CA: Rand Corporation.

______ [ed.] (1973) General Design Report: First Draft. Santa Monica, CA: Rand Corporation.

______ (1971) "Housing assistance for low-income urban families: a fresh approach," in U.S. Congress, House Committee on Banking and Currency. Papers Submitted to Subcommittee on Housing Panels, 92 Congress, 1st Session, Part 2.

______ C. P. RYDELL, and D. M. de FERRANTI (1981) Testing the Supply Response to Housing Allowances: An Experimental Design. Note N-1020-HUD, February. Santa Monica, CA: Rand Corporation.

LOWRY, I. S., B. M. WOODFILL, and T. REPNAU (1974) "Program standards for site I." Working Note WN-8574-HUD, January. Santa Monica, CA: Rand Corporation.

MALOY, C. M., J. P. MADDEN, D. W. BUDDING, and W. L. HAMILTON (1977) Administrative Costs in a Housing Allowance Program: Two-Year Costs in the Administrative Agency Experiment. Cambridge, MA: Abt Associates Inc.

MAYO, S. K. (1981) "Theory and estimation in the economics of housing demand." Journal of Urban Economics 10, 1:95-116.

——— (1977) Housing Expenditures and Quality, Part 1: Report on Housing Expenditures under a Percent of Rent Housing Allowance. Cambridge, MA: Abt Associates Inc.

——— S. MANSFIELD, W. D. WARNER, and R. ZWETCHKENBAUM (1980a) Housing Allowances and Other Rental Assistance Programs—A Comparison Based on the Housing Allowance Demand Experiment, Part 1: Participation, Housing Consumption, Location, and Satisfaction. Cambridge, MA: Abt Associates Inc.

——— (1980b) Housing Allowances and Other Rental Assistance Programs—A Comparison Based on the Housing Allowance Demand Experiment, Part 2: Costs and Efficiency. Cambridge, MA: Abt Associates Inc.

McCARTHY, K. F. (1982) "An analytical model of housing search," in W. A. V. Clark (ed.) Modelling Housing Market Search. London: Croom Helm.

——— (1980) Housing Search and Consumption Adjustment. Paper P-6473. Santa Monica, CA: Rand Corporation.

——— (1979) Housing Search and Mobility. Report R-2451-HUD. Santa Monica, CA: Rand Corporation.

McDOWELL, J. L. (1979) Housing Allowances and Housing Improvement: Early Findings. Note N-1198-HUD. Santa Monica, CA: Rand Corporation.

MacMILLAN, J. (1980) Mobility in the Housing Allowance Demand Experiment. Cambridge, MA: Abt Associates Inc.

——— and W. L. HAMILTON (1977) Outreach: Generating Applications in the Administrative Agency Experiment. Cambridge, MA: Abt Associates Inc.

MEEHAN, E. (1975) Public Housing Policy: Convention versus Reality. New Brunswick, NJ: Center for Urban Policy Research, Rutgers University.

MERRILL, S. R. (1980) Hedonic Indices as a Measure of Housing Quality. Cambridge, MA: Abt Associates Inc.

——— (1976) "The effects of segregation and discrimination on the price and quality of minority housing." Ph.D. dissertation, Boston University.

——— and C. A. JOSEPH (1980) Housing Improvements and Upgrading in the Housing Allowance Demand Experiment. Cambridge, MA: Abt Associates Inc.

MILLS, E. S. and A. SULLIVAN (1981) "Market effects," in K. L. Bradbury and A. Downs (eds.) Do Housing Allowances Work? Washington, DC: Brookings Institution.

MULFORD, J. E. (1979) Income Elasticity of Housing Demand. Report R-2449-HUD. Santa Monica, CA: Rand Corporation.

——— J. L. McDOWELL, L. HELBERS, M. P. MURRAY, and O. M. YILDIZ (1982) Housing Consumption in a Housing Allowance Program. Report R-2779-HUD. Santa Monica, CA: Rand Corporation.

MURRAY, M. P. (1975) "The distribution of tenant benefits in public housing." Econometrica 43 (July): 771-788.

MUSGRAVE, R. A. (1976) "Policies of housing support: rationale and instruments," in U.S. Department of Housing and Urban Development, Housing in the Seventies Working Papers, Vol. 1. Washington, DC: Government Printing Office.

MUTH, R. F. (1969) Cities and Housing. Chicago: University of Chicago Press.

NAPIOR, D. and A. PHIPPS (1980) Subjective Assessment of Neighborhoods in the Housing Allowance Demand Experiment. Cambridge, MA: Abt Associates, Inc.

NEELS, K. (1982a) The Economics of Rental Housing. Report R-2775-HUD. Santa Monica, CA: Rand Corporation.

______ (1982b) Revenue and Expense Accounts for Rental Properties. Note N-1704-HUD, March. Santa Monica, CA: Rand Corporation.

______ (1982c) Specification Bias in Housing Production Functions. Note N-1744-HUD, May. Santa Monica, CA: Rand Corporation.

NETZER, R. (1980) "Income strategy and housing supply," in Economics and Urban Problems. New York: Basic Books.

NEWMAN, S. J. and G. J. DUNCAN (1979) "Residential problems, dissatisfaction, and mobility." American Planning Association Journal, April: 154-166.

NOLAND, C. W. (1980) Assessing Hedonic Indices for Housing. Note N-1305-HUD, May. Santa Monica, CA: Rand Corporation.

OLSEN, E. O. (1982) "Housing programs and the forgotten taxpayer." Public Interest 66 (Winter): 97-109.

______ (1981a) "The simple analytics of the externality argument for redistribution," in M. B. Ballabon (ed.) Economic Perspectives: An Annual Survey of Economics, Vol. 2. New York: Harwood Academic Publishers.

______ (1981b) "A universal housing allowance program: comments," in K. L. Bradbury and A. Downs (eds.) Do Housing Allowances Work? Washington, DC: Brookings Institution.

______ (1969) "A competitive theory of the housing market." American Economic Review 59 (September): 612-22.

______ and D. M. BARTON (1982) "The benefits and costs of public housing in New York City." Discussion Paper 123, February. Thomas Jefferson Center for Political Economy, University of Virginia.

OLSEN, E. O. and W. J. REEDER (1981) "Are subsidies under the Section 8 Existing Housing Program excessive?" Discussion Paper 117, November. Thomas Jefferson Center for Political Economy, University of Virginia.

PEABODY, M. E., Jr. (1974) "Housing allowances." New Republic, March 9.

PETTIGREW, T. F. (1973) "Attitudes on race and housing: a social psychological view," pp. 21-84 in A. Hawley and V. Rock (eds.) Segregation in Residential Areas. Washington, DC: National Academy of Sciences.

POST, M. (1977) "Locational change in the Administrative Agency Experiment" in W. L. Holshouser, Jr. (ed.) Supportive Services in the Administrative Agency Experiment. Cambridge, MA: Abt Associates Inc.

President's Committee on Urban Housing (1968) A Decent Home. Washington, DC: Government Printing Office.

Rand Corporation (1978) Fourth Annual Report of the Housing Assistance Supply Experiment. Report R-2302-HUD. Santa Monica, CA: Author.

______ (1977) Third Annual Report of the Housing Assistance Supply Experiment. Report R-2151-HUD, February. Santa Monica, CA: Author.

RIVLIN, A. M. (1979) "How can experiments be useful?" American Economic Review 64, 2: 346-354.

RIZOR, W. E. (1982) Income Certification in an Experimental Housing Allowance Program. Note N-1740-HUD. Santa Monica, CA: Rand Corporation.

ROSSI, P. (1955) Why Families Move. New York: Free Press.

RYDELL, C. P. (1982) Price Elasticities of Housing Supply. Report R-2846-HUD. Santa Monica, CA: Rand Corporation.

______ (1980) "Supply response to the Housing Allowance Program." International Regional Science Review 5, 2: 119-138.

______ (1979a) Shortrun Response of Housing Markets to Demand Shifts. Report R-2453-HUD, September. Santa Monica, CA: Rand Corporation.

______ (1979b) Vacancy Duration and Housing Market Condition. Note N-1135-HUD, October. Santa Monica, CA: Rand Corporation.

______ and J. E. MULFORD (1982) Consumption Increases Caused by Housing Assistance Programs. Report R-2809-HUD. Santa Monica, CA: Rand Corporation.

______ and L. HELBERS (1980) Price Increases Caused by Housing Assistance Programs. Report R-2677-HUD, October. Santa Monica, CA: Rand Corporation.

RYDELL, C. P., J. E. MULFORD, and L. W. KOZIMOR (1979) "Participation rates in government transfer programs: applications to housing allowances." Management Science 25, 5: 444-453.

RYDELL, C. P., K. NEELS, and C. L. BARNETT (1982) Price Effects of a Housing Allowance Program. Report R-2720-HUD. Santa Monica, CA: Rand Corporation.

SCHNARE, A. B. (1977) Residential Segregation by Race in U.S. Metropolitan Areas. Washington, DC: Urban Institute.

SEMER, M. P., J. H. ZIMMERMAN, A. FOARD, and J. FRANTZ (1976) "A review of federal subsidized housing programs," in U.S. Department of Housing and Urban Development, Housing in the Seventies. Washington, DC: Government Printing Office.

Senate Committee on Banking, Housing and Urban Affairs (1973) Hearings on the Administrations' 1973 Housing Proposals: Anticipated Problems with Housing Allowances. Washington, DC: Government Printing Office.

SHANLEY, M. G. and C. M. HOTCHKISS (1980) The Role of Market Intermediaries in a Housing Allowance Program. Report R-2659-HUD, December. Santa Monica, CA: Rand Corporation.

SHAW, J. (1974) "Do housing allowances work?" House and Home, January, p. 8.

SILBERMAN, J. (1977) "The cost of services in the Administrative Agency Experiment," in William L. Holshouser, Jr. (ed.) Supportive Services in the Administrative Agency Experiment. Cambridge, MA: Abt Associates Inc.

SOLOMON, A. P. and C, FENTON (1974) "The nation's first experience with housing allowances: the Kansas City Demonstration." Land Economics 50 (May): 213-223.

SORENSON, A., K. TAEUBER, and L. HOLLINGSWORTH (1975) "Indexes of racial segregation for 109 cities in the United States, 1940 to 1979." Sociological Focus 8: 125-142.

STRASZHEIM, M. (1975) An Econometric Analysis of the Urban Housing Market. New York: National Bureau of Economic Research.

STRUYK, R. J. and M. BENDICK, Jr. [eds.] (1981) Housing Vouchers for the Poor. Washington, DC: Urban Institute.

TAEUBER, K. E. and A. F. TAEUBER (1965) Negroes in Cities: Residential Segregation and Neighborhood Change. Chicago: Aldine.

TAYLOR, D. G. (1979) "Housing, neighborhoods, and race relations." Annals of the American Academy of Political and Social Science 441: 26-40.

TEBBETS, P. E. (1979) Controlling Errors in Program Administration. Note N-1145-HUD, August. Santa Monica, CA: Rand Corporation.

TEMPLE, F. T. and R. WARLAND (1977) "Factors associated with enrollees' success in becoming recipients," in W. L. Holshouser, Jr. (ed.) Supportive Services in the Administrative Agency Experiment. Cambridge, MA: Abt Associates Inc.

TREND, M. G. (1977) "Services in the Administrative Agency Experiment: a discussion of three different approaches," in W. L. Holshouser, Jr. (ed.) Supportive Services in the Administrative Agency Experiment. Cambridge, MA: Abt Associates Inc.

U.S. Comptroller General's Office (1974) Observations on Housing Allowances and the Experimental Housing Allowance Program. Washington, DC: Government Printing Office.

U.S. Congress, Joint Economic Committee (1972) Housing Subsidies and Housing Policies, Hearings before the Subcommittee on Priorities and Economy in Government, 4, 5, and 7 December.

U.S. Department of Housing and Urban Development (1980) The Experimental Housing Allowance Program: Conclusions, The 1980 Report. Washington, DC: Government Printing Office.

______ (1973) Housing in the Seventies. Washington, DC: Government Printing Office.

U.S. Department of Labor, Bureau of Labor Statistics (1972-1977) CPI Detailed Reports. Washington, DC: Government Printing Office.

U.S. General Accounting Office (1973a) An Analysis of the Section 235 and 236 Programs. Washington, DC: Government Printing Office.

______ (1973b) Social Services: Do They Help Welfare Recipients Achieve Self-Support or Reduced Dependency? Washington, DC: Government Printing Office.

VALENZA, J. J. (1977) Program Housing Standards in the Experimental Housing Allowance Program: Analyzing Differences in the Demand and Supply Experiments. Washington, DC: Urban Institute.

VANSKI, J. E. and L. J. OZANNE (1978) Simulating the Housing Allowance Program in Green Bay and South Bend. Washington, DC: Urban Institute.

VIDAL, A. (1980) The Search Behavior of Black Households in Pittsburgh in the Housing Allowance Demand Experiment. Cambridge, MA: Abt Associates Inc.

WALLACE, J. E., S. P. BLOOM, W. L. HOLSHOUSER, S. MANSFIELD, and D. H. WEINBERG (1981) Participation and Benefits in the Urban Section 8 Program: New Construction and Existing Housing. Cambridge, MA: Abt Associates Inc.

WATTS, H. W. with F. SKIDMORE (1981) "A critical review of the program as a social experiment," in K. L. Bradbury and A. Downs (eds.) Do Housing Allowances Work? Washington, DC: Brookings Institution.

WEAVER, R. C. (1975) "Housing allowances." Land Economics 51, 3: 247-257.

WEICHER, J. C. (1979) "Urban housing policy," in P. Mieskowski and M. Straszheim (eds.) Current Issues in Urban Economics. Baltimore: Johns Hopkins University Press.

______ (1976) "Policy and economic dimensions of 'A Decent Home and a Suitable Living Environment.'" Presented to the American Real Estate and Urban Economic Association, May.

______ (1973) "The rationales for government intervention in housing: an overview," in U.S. Department of Housing and Urban Development, Housing in the Seventies. Washington, DC: Government Printing Office.

WEINBERG, D. H. (1982) "Housing benefits from the Section 8 Housing Program." Evaluation Review 6 (February): 5-24.

______ J. FRIEDMAN, and S. K. MAYO (1981) "Intraurban residential mobility: the role of transactions costs, market imperfections, and household disequilibrium." Journal of Urban Economics 9: 332-348.

WENDT, J. C. (1982) The Decision to Apply for a Housing Allowance. Report R-2782-HUD. Santa Monica, CA: Rand Corporation.

______ (1981) Why Households Apply for Housing Allowances. Report R-2783-HUD, May. Santa Monica, CA: Rand Corporation.

WHITE, E. S., S. R. MERRILL, and T. LANE (1979) The History and Overview of the Performance Funding System. Cambridge, MA: Abt Associates Inc.

WIENK, R. F., C. E. REID, J. C. SIMONSON, and F. J. EGGERS (1979) Measuring Racial Discrimination in American Housing Markets: The Housing Market Practices Survey. Washington, DC: U.S. Department of Housing and Urban Development, Office of Policy Development and Research.

WILD, B. (1977a) "Residential movement among recipients in the Administrative Agency Experiment: an analysis of moving intentions, housing search, and active moves," in W. L. Holshouser, Jr. (ed.) Supportive Services in the Administrative Agency Experiment. Cambridge, MA: Abt Associates Inc.

______ (1977b) "The effects of agency services on enrollees' success in becoming recipients," in W. L. Holshouser, Jr. (ed.) Supportive Services in the Administrative Agency Experiment. Cambridge, MA: Abt Associates Inc.

WOLFE, M. F. and W. L. HAMILTON (1977) Jacksonville: Administering a Housing Allowance in a Difficult Environment. Cambridge, MA: Abt Associates Inc.

About The Contributors

☐ *C. LANCE BARNETT* is a senior economist and policy analyst with the Rand Corporation. Since joining Rand in 1975, he has concentrated on studying how housing markets function.

☐ *MARC BENDICK, Jr.* is an economist specializing in problems of poverty, income transfers, and the management of social assistance programs. He is now a senior research associate at the Urban Institute, where he was manager of the institute's work on the Experimental Housing Allowance Program from 1977 through 1979. He is coauthor of *Housing Vouchers for the Poor.*

☐ *GRACE M. CARTER* is with the Rand Corporation. Her current research concerns analysis of military personnel policies, biomedical research manpower, and federal policies for the support of biomedical research.

☐ *SINCLAIR B. COLEMAN* is an independent consultant and teacher. His research areas include health policy, housing, demographic modeling, and defense analyses.

☐ *JOSEPH FRIEDMAN* is a member of the Economics Department and the Center for Urban Studies at Tel Aviv University in Israel. He is a coauthor of *The Economics of Housing Vouchers* (1982).

☐ *WILLIAM L. HAMILTON* is a Vice President of Abt Associates, where he directed the Administrative Agency component of the housing allowance experiments as well as participating in the analysis of the Demand Experiment. He is the author of *A Social Experiment on Program Administration.* His current work focuses on analysis of administrative reforms in welfare programs.

□ *CAROL E. HILLESTAD* is an assistant social scientist for the Rand Corporation in the field of housing. Her research work includes neighborhood change, rent control, and household housing consumption. She is currently analyzing benefit-based financing for local governments.

□ *WILLIAM L. HOLSHOUSER, Jr.* is a social scientist with Abt Associates Inc. His current research includes an analysis of the fair housing assistance program and an analysis of the impacts of monthly reporting systems on welfare clients.

□ *STEPHEN D. KENNEDY* is a Vice President of Abt Associates, where he directed the Housing Allowance Demand Experiment. His current work involves hospital cost containment and criminal justice research.

□ *IRA S. LOWRY* is at the Rand Corporation, where his current research concerns housing policy for California. His most recent articles include "The Dismal Future of Central Cities" and "Rental Housing in the 1970s: Searching for the Crisis." Dr. Lowry was the principal investigator of the Housing Assistance Supply Experiment. The final report of that experiment, *Experimenting with Housing Allowances*, has recently been published.

□ *KEVIN F. McCARTHY* is a social demographer at the Rand Corporation, where he worked on the Housing Assistance Supply Experiment. His research interests center on the determinants and consequences of demographic trends.

□ *JAMES L. McDOWELL* is the Planning Systems Manager for General Telephone of California. He directs development of a mixed integer program to optimize corporate capital investment decisions and manages companywide evaluation activities.

□ *JEAN E. MacMILLAN* is studying for her doctorate in cognitive psychology at Harvard University. She has spent many years at Abt Associates working on social welfare problems.

□ *STEPHEN K. MAYO* is an economist with the World Bank, where he specializes in the urban and housing problems of developing countries. His recent publications include "Theory and Estimation in the Economics of Housing Demand."

☐ *SALLY R. MERRILL* conducts policy research at Abt Associates Inc. Her current interests include housing and the elderly, public housing, and quality change in the housing stock.

☐ *JOHN E. MULFORD* is associate professor of economics in the School of Business Administration at CBN University. Previously, he was vice president and senior economist at First Interstate Bank of California. He conducted this research while working as a housing economist at the Rand Corporation.

☐ *EDGAR O. OLSEN* is a Visiting Professor of Economics at the University of Wisconsin, Madison on leave from the University of Virginia. He has been involved for more than fifteen years in developing methods for housing policy analysis and applying them to major government programs such as rent control, public housing, and Section 8.

☐ *C. PETER RYDELL* does public policy research at the Rand Corporation. He also teaches in the Rand Graduate Institute and the UCLA Economics Department. Recent publications include *Price Elasticities of Housing Supply* and *The Impact of Rent Control on the Los Angeles Housing Market.*

☐ *RAYMOND J. STRUYK* is at the Urban Institute, where he is currently Director of the Center of International Activities. His publications include *A New System for Public Housing, Improving the Elderly's Housing,* and *Housing Vouchers for the Poor* (coauthored with Marc Bendick). Dr. Struyk is a former Deputy Assistant Secretary for Research in the U.S. Department of Housing and Urban Development.

☐ *DANIEL H. WEINBERG* is an economist with the Office of Income Security Policy in the U.S. Department of Health and Human Services. His research interests include the economics of transfer programs and poverty, income distribution, and taxation. His recent publications include "The Housing Benefits of the Section 8 Housing Program," and he is coauthor of *The Economics of Housing Vouchers.*

☐ *JAMES C. WENDT* is a policy analyst at the Rand Corporation, where he is currently working on issues relating to national security.